STONE ROOFING

Conserving the materials and practice of traditional stone slate roofing in England

ENGLISH HERITAGE RESEARCH TRANSACTIONS

RESEARCH AND CASE STUDIES IN ARCHITECTURAL CONSERVATION

STONE ROOFING

Conserving the materials and practice of traditional stone slate roofing in England

EDITED BY

Chris Wood

Volume ***9***

August 2003

Published with the assistance of the Getty Grant Program.

Published by James & James (Science Publishers) Ltd,
8–12 Camden High Street, London NW1 0JH, UK

A catalogue record for this book is available from the British Library
ISBN 1-902916-32-8
ISSN 1461 8613

Volume editor: Chris Wood, English Heritage
Series editor: John Stewart, English Heritage
Consultant editor: Kate Macdonald

Printed in the UK by Stones the Printers

Front cover

Hipped stone roof of Stokesay Castle, Shropshire
(© English Heritage Photo Library)

Contents

Acknowledgements

A great many people have contributed to the projects described within this work and they are cited below under the relevant chapter headings. Particular thanks should be made to those who were instrumental in bringing this volume to fruition. The geological acumen, advice and positive intervention of Dr David Jefferson were fundamental to all three studies presented here. John Fidler, Director of Conservation at English Heritage (and former Head of its Building Conservation & Research Team), is to be thanked for his sustained support for these projects, and Dr Kate Macdonald, consultant editor, for her guidance over the long gestation period before publication. The Getty Grant Program provided a very generous grant which enabled all photographs in this volume to be reproduced in colour. We are most grateful to them, as colour images alone can show the range and distinctive diversity of stone roofing in England.

SAVING ENGLAND'S STONE SLATE ROOFS

The authors are greatly indebted to the many people of Derbyshire and the Peak Park who contributed to the research into the history, geology and state of the stone slate industry in the South Pennines. They were consistently generous with their time and knowledge. Special thanks are due to Barry Joyce and Alan Morrison of Derbyshire County Council's Conservation and Design Group who initiated the stone roofing work and whose enthusiasm and determination to resolve the problem of conserving the region's roofs has been sustained over several years and in the face of many setbacks. They were also supported by Mike Lea, formerly conservation officer of the Peak District National Park Authority.

Ian Thomas made a major contribution to the field work through his understanding of the complex geology of the region. Without this extensive knowledge and the help of his staff at the National Stone Centre the research would have been very much more difficult.

Oxford Brookes University has supported the work to conserve stone roofs through both the parallel petrological study of the South Pennine stone slates and by hosting the original stone roofing website. We are very grateful to Paul Guion who led the geological study and managed the website.

The results of these studies were publicised in a national briefing at De Montfort University in 1998. Thanks are due to Dr David Watt for ensuring everything ran smoothly.

STONE ROOFING IN ENGLAND

Many people helped in compiling this report. Special thanks are due to Dr David Jefferson. This report is more accurate because of his help, but any mistakes which it may contain are the author's. Dr Graham Lott of the British Geological Survey generously provided advice and information on the more abstruse aspects of the building stone industry in the UK and its geology. The author is also grateful to Paul Guion and the Geology staff at Oxford Brookes University for their work in characterising sandstone slates. For helpful comments and for allowing access to their research, the author is also grateful to Dr Ian West (Portland and Purbeck), Jo Thomas (Dorset and Somerset), Dan Martin (UK), Jeremy Price (Cotswolds), Dr Diana Sutherland (East Midlands), Lisa Brooks (Horsham), Dr A Hughes (Horsham), Ian Thomas (Derbyshire and the Peak Park), Arthur Baldwin (Rossendale) and the Yorkshire Dales National Park. Thanks are also due to many colleagues in the field of stone roof conservation especially Susan Macdonald, Chris Wood, Gerald Emerton, Alan Morrison, Chris Harris and John Wheatley, who were generous with their help and support.

PITCHFORD

Above all the re-roofing of Pitchford Church was a team effort. Without the participants' enthusiasm and their confidence that the objectives would ultimately be achieved in the face of seemingly overwhelming problems it might have floundered on many occasions. The authors' gratitude is due to Alex Argyropolu of the Parochial Church Council for liaising with the congregation and for other help in ensuring a successful outcome.

Thanks are also due to Dr David Jefferson for help with the geological survey and for intervening at short notice to assist with the planning application. Malcolm Bell of Shropshire County Council Planning Department was also unfailingly helpful in guiding the smooth progress of the application.

Recognition also needs to be given to David Heath and Trudi Hughes (formally Architects in the West Midlands Region of English Heritage), and to Susan Macdonald whose enthusiasm in the early 1990s to reroof the nave in Harnage slate, and whose resistance to using another material, were instrumental in initiating the project.

For permission to carry out the trial excavations we are grateful to Mr J Wilde of Bull Farm and the Trustees of the Acton Burnell Estate. The Trustees also generously allowed the delving operation. For this, thanks are due to Sue Baker of Balfour and Cooke Land Agents.

Andrew Arrol, the Parochial Church Council's architect, and John Wheatley, English Heritage's commissioned architect, always took a practical approach in dealing with a difficult product and finding ways of achieving an appropriate conservative repair as well as a technically functional roof. Andrew Arrol also kindly permitted reproduction of his roofing drawings.

Thanks are also due to the two quantity surveyors, Wilf Jones (acting for the PCC) and Ian Forrest of English Heritage, who monitored output during the delving and maintained tight control of expenditure over the whole project.

Making the Harnage stone into usable stone slates and assembling them into an effective roof was never going to be easy. The dedication of Chris Harris and his team in producing the slates must be acknowledged; especially Ivor Curnick, Neil Plumb and Martin Hillier who quarried the stone in appalling weather. Dave Rowlands of Heritage Roofing was consistently committed to building a good roof even when the odds were against it.

Introduction

Chris Wood
English Heritage, 23 Savile Row, London W1S 2ET, UK. Tel: +44 20 7973 3026;
Fax: +44 20 7973 3130; email: chris.wood@english-heritage.org.uk

BACKGROUND

Stone roofs give old buildings a most endearing and distinctive quality, especially where lichens and small mosses have mellowed their appearance over time. Most of the different types of stone roofs found in England are either of limestone or sandstone. 'Limestone can yield a roof covering of incomparable dignity and beauty. When the building is itself constructed of local limestone, a roof of stone slates adds the crowning touch of harmony, in colour and in texture, with the surrounding landscape' (Clifton-Taylor 1962). Similar sentiments can be attributed to sandstone.

The problem is that most of these unique stones are no longer quarried and many of these attractive roofs have disappeared over the last fifty years. This trend is continuing. English Heritage has been trying to reverse this over the last few years and their campaign aims to highlight the issues and publish information which could prove useful to those who want to produce new stone slates. This volume describes three initiatives that formed a major part of the campaign, *The Roofs of England* (1997–99).

Rejuvenating an industry, or more particularly a series of small industries, parts of which have been dormant for over a century, is a daunting prospect and not one that will be realised fully. Indeed, one of the studies in this volume concludes that efforts should be concentrated on the most significant and distinctive types of stone slates, which could be used to fill gaps elsewhere. However, wherever there is local enthusiasm to provide a supply of new material, encouragement will be given. New delves (small quarries) are being opened and this volume provides advice and information to those intending to embark on such a project.

The indications so far are promising, but it is clear that our predecessors had already found most of the sources of fissile or naturally splittable rock suitable for roofing. The case study in Shropshire shows that delving for stone slates can be highly sustainable: it is low-impact, mainly handcrafted, often short-term and serves local needs. It needs encouragement and support. Although some of these old delves still have an exploitable supply, they also make excellent habitats for flora and fauna and are now often protected for archaeological or nature conservation reasons. This volume aims to raise the profile and importance of stone slate roofing and to demonstrate that a sensible balance between these competing objectives is needed, if we are to avoid a serious erosion of our historic buildings and roofscapes.

A note needs to be added on the use of the words 'delph' and 'delving', used extensively in this volume as synonyms for 'quarry' and 'quarrying'. Both are old words for quarry, and reflect local usage. They are preferred in relation to the production of stone slates because they do not carry the negative connotation of large-scale activity that 'quarry' and 'quarrying' do. In Purbeck the traditional noun is 'quarr'.

STONE ROOFING

Stone was the favourite roofing material for many parts of England from Tudor times until the first quarter of the nineteenth century. This produced very distinctive buildings and roofscapes which owed much to the rich variety of local stones. The delves produced slates of very different colours, textures, thicknesses and sizes, which in turn determined the appearance and detailing of roofs: so important a feature of the local distinctiveness of our regions. Cotswold stone roofs are distinguished by their steep pitches, swept valleys and ornate dormers, whereas the larger and thicker sandstone slates of the Pennines are more suited to shallower pitches and simpler designs creating a solid, four-square style. Different minerals in the stone and the aspect of a roof also determined the lichens and other plants that would colonise it; providing further subtleties in colour and appearance.

The advent of canals and railways in the nineteenth century meant that cheaper Welsh slates and Midlands clay tiles drastically reduced demand for stone slates. Production was further affected in the twentieth century by the loss of experienced delvers (quarrymen) through the two world wars and the increasing availability of concrete tiles and artificial slates.

In the last thirty years sustained efforts to conserve historic buildings have meant that second-hand stone slates have been sought widely, often with the help of grants. As a result, the roofs of unlisted buildings have been cannibalised which has greatly impoverished the appearance of many historic roofs and landscapes. With very little new supply (at the beginning of the 1990s the only sources were the Cotswolds or Yorkshire), imported material, particularly from France and, latterly, India, was being used increasingly. While these may have initially resembled the original materials, weathering characteristics and subtle detailing requirements have often meant that they do not produce an authentic appearance.

This problem was particularly acute in Derbyshire and the Peak Park where a variety of stone slates was evident on roofs in the county, but only one was available, and this was inappropriate for most of the region. A seminar was held in 1993 to look at ways of rejuvenating the industry and this resulted in a joint initiative being established between the County Council, the Peak Park Joint Planning Board and English Heritage. Terry Hughes was commissioned to carry out a detailed study of the problem in the South Pennines to be used as a model for others wanting to revive their dormant stone roofing industries. Two volumes were produced in 1996 and the whole exercise is summarised in the first paper in this book, 'Saving England's stone slate roofs'.

The South Pennines study looked at the factors that would determine whether the industry could be revived. The starting point was to understand clearly the history and evolution of stone slating in the area, and Pat Strange's comprehensive work clearly demonstrated its importance in contributing to the unique character of Derbyshire. Unlike many other areas of England, the industry survived in Derbyshire until the 1950s, but forty years later it had disappeared. The study took a pragmatic look at the chances of reviving production. The potential market for new stone was investigated, including how this could be supported and the products promoted, together with the implications of prevailing planning policies, particularly minerals planning. Other key factors included suggestions for overcoming difficulties with the supply chain, the provision of product advice, training for specifiers, craftsmen and tradesmen, carrying out environmental assessments and the criteria for conserving historic roofs.

The study highlighted potential delves and proposed seven generic stone types which would reasonably cover the traditional variety in the region. It also concluded that the market would be fairly limited, production would probably be small-scale and still mainly hand-crafted. Improved machinery might help initial extraction but hand-working would still be needed for splitting and dressing stone slates if their aesthetic appeal was to remain. Profit margins would therefore be very tight and as stone suitable for splitting would often be found with less fissile rock, it was clear that production of stone for other uses such as flags, pavers and walls could be essential if a delph was to be viable.

Historical research showed that most of the potential sources of stone had already been exploited, although some reserves remained. Different scales of delving operation were proposed. These ranged from steady production involving a few lorry loads per week, to a 'mobile operation' which would merely open a small delph for a few weeks until enough material was won to roof a particular building. The study clearly showed that winning stone slates is a very different proposition in scale, duration and environmental impact than is normally associated with quarrying. This was a message that needed to be spread widely if the industry stood any hope of being revived and the political tensions that the word 'quarrying' usually invokes were to be allayed.

In 1996 English Heritage began a nationwide campaign, 'The Roofs of England', in order to publicise the results of the South Pennines study and galvanise interest in the other stone slate areas. Two exhibitions and various leaflets were produced and the messages were disseminated at conferences, seminars and in the media. One of the most important documents produced was English Heritage's technical advice note, *Stone Slate Roofing* (1998) which sought to help specifiers, craftsmen, conservation officers and others on detailed aspects of repair and reroofing. Important principles were proposed, including the unequivocal recommendation that new stone slates should be used in preference to second-hand material taken from other buildings (good conservation principles dictate that sound reusable slates from an existing building should be used again *on the same building*). Another very important section of the advice note was the inclusion of a check-list for recording construction details. It was apparent from the South Pennines study that, with few roofers working on traditional stone roofs and no skilled men delving the product, much essential information would otherwise be lost. Elsewhere, where no new slates had been produced for over a century it was quite likely that new 'traditions' had evolved. Information on regional variations was and still is, urgently needed.

The campaign revealed that many regions of England only had fragmentary information on their stone slating traditions and the potential demand for new supplies. Recognition of this led English Heritage to commission Terry Hughes to carry out his second study, 'Stone roofing in England', which is the second paper in this volume. The work began in October 1998 and concluded in 2000, but has been updated to the end of 2001. As well as endeavouring to assess supply and demand for each main stone slate region, he has also brought together published, unpublished and anecdotal material on the sources of stone slates for the first time. This information is intended to be made as widely available as possible, so the photographic images will be available on www.stoneroof.org.uk with a link to the English Heritage website.

Terry Hughes' research shows just what a rich tradition of stone slate roofs England enjoys. Essentially, the number of different types is almost limitless, because the vagaries of their geological formation mean that significant variations in colour, texture and mineral composition are found within a region or even in the same delph. The study makes hard-headed conclusions about the efficacy of trying to revive the whole industry. It would simply not be realistic to attempt this, because of the vast range of stone types, limited demand, marginal profits, exhausted supplies or overwhelming environmental constraints. To decide what is practical and achievable, decisions need to be made about which stone slates are so distinctive that every effort should be made to revive them. For those that are not especially distinctive and where it is unrealistic to try to exploit their original sources, a clear definition of their geological and visual characteristics is needed so that matching can take place.

This, together with performance and durability tests, can provide some form of validation for a 'new' product used in a different area. 'Stone roofing in England' reiterates the urgent need to study old roofs to learn about traditional detailing in different regions. New technical improvements for productivity and to reduce costs are also desirable, although these need to be evaluated carefully to ensure that the appearance and durability of the slates are not unduly compromised.

The study has shown that there are grounds for optimism in the future. With a certain amount of business acumen and determination, farmers and entrepreneurs have reopened long dormant sources, providing Pennant, Harnage and Herefordshire sandstones and new limestone sources in the Cotswolds. Planning permission has been granted for two delves in the Peak Park and other applications have been made there and in Derbyshire. There is no technical reason why Collyweston slate and Horsham stone should not come into production in the near future.

The third paper in this volume, 'Sourcing new stone slates and re-roofing the nave of Pitchford Church, Shropshire', describes the winning of Harnage stone in 1997 which allowed the church to be re-slated the following year. This project provided a unique opportunity to test many of the principles and recommendations that had come out of the studies and the English Heritage technical advice note. Its importance lay in the many lessons that were learned from all the pitfalls of finding and winning a source of stone that had not been produced commercially for over a century. As well as the technical issues involved, it must be stated that the stone roof on the church had failed after only sixty years and there was much pressure from some members of the Parochial Church Council and their advisors to change to machine-made tiles. Consequently, this case also became a test to demonstrate that a source of stone could be found, successfully exploited and the faults with the existing roof diagnosed and rectified in the new design.

One of the most instructive lessons came from the documentary research and detailed investigation into the roof and its repair over the last century. It became apparent that the 1930s repair had failed because of the inability of the suppliers to obtain stone that could provide adequately sized slates. The contractor had tried to rectify this by using mortar to bed the slates with the result that water had entered, become trapped and rotted the laths and pegs. Understanding the principles of random stone roofs is an essential prerequisite to the investigation of a roof and the appropriate design of the new roof. Effective research of sources of stone and close liaison with landowners and relevant authorities are also vital if the requisite material is to be won. Even then, there is still an element of risk and tension because, until the stone is actually quarried, it is impossible to be certain about its quality, the proportion of useable rock or the costs.

A number of shortcomings within the project are discussed, but it is often from mistakes that lessons are best learned. The importance of involving experienced roofers familiar with the stone and detailing cannot be stressed enough. Technological improvements provided by the supplier, in this case using computer information to deliver just enough slates for each course, does not work as it does not allow the roofer the scope to adjust his setting out to ensure adequate side laps. Changing the traditional detail from pegging on laths to nailing onto boarding with protective felts might imply a lack of confidence. In this instance it was motivated by a need to ensure that the chronic condensation problem would not cause continual staining to the painted internal roofs, as well as increasing the comfort of the congregation.

Perhaps the most significant plus factor to come out of this work was that the whole delving operation took only eight weeks and while it looked like an ugly scar during the works it was quickly restored afterwards. The stone produced enough slates to re-roof the nave of the church (with a supply for repairs and maintenance) and for three other historic buildings in the area. We also now know where a supply of good quality material exists for future requirements.

Each of the three papers in this volume highlight further work that needs to be done, if there is to be a real revival of the industry. English Heritage helped to establish the Stone Roof Working Group, comprising delph operators, roofing contractors, trainers and conservation & minerals planning officers and roofing consultants who have produced a detailed Guidance Note for applicants wishing to apply for permission to open a delph for stone slates. A 'best practice' note has been prepared covering the conservation of stone roofs. In a separate initiative, Gerald Emerton, one of the members, is preparing the NVQ slating and tiling training manuals for the Construction Industry Training Board. This includes random roofing at level 3.

REFERENCE

Clifton Taylor A, 1962 *The Pattern of English Building*, London, Faber.

Saving England's stone slate roofs

A model for the revival and enhancement of the stone slate roofing industry in the South Pennines

SUSAN MACDONALD
formerly of English Heritage, Building Conservation & Research Team, 23 Savile Row, London W1S 2ET, UK.
TERRY HUGHES
Slate and Stone Consultants, Ceunant, Caernarfon, Gwynedd LL55 4SA, UK. Tel: +44 1286 650402;
email: terry@slateroof.co.uk
CHRIS WOOD*
English Heritage, Building Conservation & Research Team, 23 Savile Row, London W1S 2ET, UK.
Tel: +44 20 7973 3026; fax: +44 20 7973 3130; email: chris.wood@english-heritage.org.uk
PATRICK STRANGE
South of Ivy Bank, Church Street, Brassington, Matlock, Derbyshire DE4 4HJ, UK. email: strangepatrick@aol.com

Abstract

Stone slates create a highly regionalised roofing form that is fundamental to the distinctive local character of vernacular buildings in many parts of the country. Declining supplies of new stone slates and the use of imported or artificial substitutes have gradually eroded this vernacular roofing tradition, placing these distinctive landscapes under threat. Meanwhile, the conservation requirement that repairs to historic stone slate roofs on listed buildings be carried out on a like-for-like basis, has fuelled the market for salvaged stone slates and, in some cases, encouraged the unnecessary and unscrupulous stripping of other historic buildings in the locality.

Recognising that action was needed to secure the future of the stone slate roofing tradition, English Heritage joined forces with the Peak Park Joint Planning Board and Derbyshire County Council to carry out research into the revival and enhancement of the stone slate roofing industry in the South Pennines. The research was intended to act as a model which could be applied to other regions with a stone slate tradition, or which faced similar problems with other scarce historic building materials.

This chapter describes the work carried out to date and makes recommendations for the future.

Key words

Stone slate roofing, building traditions, quarrying, building industry, planning

INTRODUCTION

> If any regard is to be had to the general beauty of the landscape, the natural material of the special countryside should be used instead of imported material. (Morris, 1890)

Stone slates have been used for roofing in Britain since Roman times. Originally a prestigious alternative to thatch, they eventually became a standard material for roofs in many parts of the country. The richness of our national roofscape stems mainly from the great variety of British geology. Wherever a rock could be split to form a reasonably thin slab, it has been exploited for roofing, and examples exist from almost every geological period and rock type.

Until the second half of this century, stone slates were mined or quarried on a relatively small scale in many places around Britain. The following list gives the principal areas of the country which sustain a stone slate roofing tradition and is a reminder of how extensively stone slate roofing once was (slate types are given in italics).

- Dorset: *Purbeck, Forest Marble*
- south and east Somerset: *Ham Hill, Forest Marble*
- north Wiltshire, Gloucestershire, Oxfordshire: *Purbeck, Corallian, Forest Marble, Hampen Marly Beds, Taynton* and *Trougham tilestones* including *Stonesfield Slate, Chipping Norton Limestone, Lower Fullers Earth, Inferior Oolite*
- Surrey, Sussex, Kent: *Horsham stone, Charlwood Slate*
- Gloucestershire, south Wales: *Carboniferous Pennant Sandstone, Tilestones, Old Red Sandstone, Lower Lias?*
- Welsh Marches: *Old Red Sandstone, Tilestones, Cheney Longville Flags, alternata Limestone, Chatwall stone, Harnage Slates, Corndon Hill dolorite*
- Staffordshire, Cheshire, Derbyshire, Lancashire, Yorkshire, County Durham, Northumberland: *Carboniferous Millstone Grit and Coal Measures sandstones*
- East Derbyshire, Nottinghamshire: *Magnesian Limestone*
- Cumbria: *Carboniferous Millstone Grit* and *Coal Measures sandstones, Permian Sandstone*
- Northamptonshire, Rutland and the Kesteven region of Lincolnshire: *Collyweston Slate, Duston stone*
- North Yorkshire: *Brandsby limestone, Boltby Slate*

Stone slating is thus a highly regionalised roofing form, fundamental to the distinctive local character of vernacular buildings in many parts of the country. Its solid beauty achieves a visual harmony with the stone buildings and

* Author for correspondence

the fields' drystone walls and makes such areas as the Cotswold or the Pennines unique. However, this distinctive local character is under threat as sandstone and limestone roofing becomes increasingly rare. As locally produced stone slates come to the end of their natural lives, declining supplies have led to their replacement by imported or artificial substitutes which have little to do with local vernacular traditions.

The catalyst for this cycle of substitution and loss of local character was the development of transport systems which enabled cheaper products to be imported from other regions of Britain. Today, we can see the same process on an international scale. The consequence has been that the production of many stone slate types ceased years ago and the suppliers of the few types that are still available are in a precarious situation, largely dependent on indirect support through the grant-aiding heritage bodies for their continued existence.

Stone slates are randomly sized and all stone roofs are laid to diminishing courses, arranged with the largest slates at the eaves and reducing evenly and regularly to the smallest at the ridge. This was to maximize the output from the quarried rock. They may vary from 3 feet or even 4 feet long down to 4 inches long (0.914 m or 1.21 m long, down to 100 mm), although few roofs would encompass such a large range.

More commonly, roofs would be built from a smaller range of sizes, either occurring naturally from a particular quarry or as a result of selection into two or more length ranges. Larger sizes were preferred for their economy of labour and superior weather-proofing ability. There might also be a choice between stones with flat split surfaces and the rougher products. In practice, however, the difficulties of transport meant that whatever material could be obtained locally was used in preference.

The skill of the slater and his style of working added the final dimension that gave the local and regional character to stone roofs. Distinctiveness derived from the colour and texture of the rock; the size, thickness, flatness and surface texture of the slates; and the treatment of hips, valleys, ridges, dormers, eaves and abutments. All were intended primarily to resist the weather but also allowed the expression of the craftsman's artistry.

The South Pennines is typical of many areas of the country struggling to retain a stone slate roofing tradition, using a variety of sandstone slates. The primary influences on their appearance were the conditions in which the beds of sand were deposited when the rock was being formed. Sediments laid down in shallow water might be rippled or braided and these features persist in the roofing slate. In deeper, calmer waters, they would be flat, producing smooth, regular and generally larger slates. In slower-moving water, the sand particles would be smaller and so the slate's texture would be finer and the surface less grainy.

Add to this a variety of colours, which ranged from pale yellow to red, all produced by iron minerals, and the regional distinctiveness becomes truly local, changing even from village to village. Difficult terrain ensured that each community used its local stones, until improvements in roads and then the canals and railways made it easy to move products to more distant markets. At first, competition was local, between one stone slate and another, but gradually industrial development in the clay tiles of Staffordshire, the metamorphic slate quarries of Wales and, more recently, the ubiquitous concrete tile had an increasing influence until, eventually, stone slate production in the South Pennines ceased altogether.

Since then, the repair and renewal of old stone slate roofs and the construction of new ones has depended on supplies of second-hand slates or new ones imported from outside the region. Neither of these sources are satisfactory: the former leads to the cannibalisation of other roofs, often unnecessarily and at worst by theft; the latter rarely match the originals.

The conservation movement's response to these difficulties has evolved over the years and involved planning controls and grants to help ameliorate the financial consequences of the control. Although helpful, this approach did not encourage the supply of new slates. Even worse, the grant system supported the use of reclaimed slates in a way that led to the further decline of the regional roofscape by encouraging the cannibalisation of other roofs. Architectural and conservation professionals were ill-equipped to specify and control the use of the most appropriate products and consequently roofs were often constructed or repaired using unsuitable techniques.

It was also clear that these problems were not restricted to the South Pennine region, as the same situation exists wherever stone slates were traditionally used as a roof covering. The problem is also not restricted to stone slates but is shared with many other traditional products used on historic buildings. In an age when standardisation and mechanisation are seen as economically necessary, it is increasingly difficult for many traditional materials to compete with synthetic and mass-produced substitutes, despite the many advantages they may offer.

Conservation is about understanding what makes buildings and areas distinctive, and determining what balanced package of affordable measures may be required, now and in the future, to manage change and preserve these special qualities. To be successful, therefore, it must form an integral part of the social, economic, environmental and cultural development of the region. Based on this principle, English Heritage, the Peak Park Joint Planning Board and Derbyshire County Council joined forces to carry out research into the issues surrounding stone slate roofing in order to retain and develop this modest indigenous industry in the South Pennines. This chapter provides an outline of the work up to 1999. It is intended that this research will provide a model which could be applied in the future to solve similar regional problems in the other areas of the country struggling to sustain a stone slate tradition.

PHASE I – Devising a model

Research aims

In April 1993, a seminar on the problems of stone slate roofing in the South Pennines was organised by Derbyshire County Council, bringing together the planning bodies, quarry owners, roofers and specifiers in an attempt to define the problem and find a solution. The clear conclusion emerged that there was a need for research into the size of the market; the impact of policies for mineral planning, development control and historic building grant aid; educational and training needs; sources of financial support; and the provision of information to specifiers, manufacturers and users.

It was against this background that the present study was established. Supported by English Heritage, Derbyshire County Council and The Peak Park Joint Planning Board its objectives were to:

- review the state of the industry: the reasons why production had ceased and the potential for its re-establishment
- review the market for stone slates, the sources of competition, restraints on its development and sources of support, including grant aid
- review the relevant mineral and conservation planning policies to determine how present policies were controlling or contributing to the problem
- review the training needs of manufacturers, roofing contractors, architects and planning and conservation officers, in order to support and control the use of new and reclaimed stone slates
- produce a database of historical sources of stone slates in the region, including a description of the slates and a photographic record
- develop a methodology which would be applicable to investigations of stone slate and similar traditional building products in other regions.

As part of this study, a report was commissioned on the history of vernacular architecture in Derbyshire, with particular reference to the use of materials.[1] This proved extremely helpful in understanding the use of different roofing materials and determining patterns of use and distribution over the centuries. The study was carried out over a year from November 1994 and was managed by a steering group comprising representatives of the three supporting organisations.

It was recognised at the outset that there would be insufficient funds to carry out a full geological study of the source rocks. This potentially serious gap was filled by the Construction Products & Materials Sponsorship Division of the Department of the Environment which grant-aided a parallel geological study. This study was carried out by T G Hughes of Slate and Stone Consultants, I A Thomas of the National Stone Centre, Wirksworth, and Dr P D Guion, A M O'Beirne and Dr G R Watt of the Department of Geology and Cartography at Oxford Brookes University. The results were reported in *Roofing Stones in the South Pennines: A Geological Study of the Origins, Sources and Suitability of Sandstones for Roofing* (Hughes et al 1995).

In order to address all the issues which have led to the decline of the industry and the current threat to the historic character of the South Pennines, the study had to tackle the interrelated issues which successful conservation must address if it is to be effective. This chapter is a shortened version of the research report for Phase I (Hughes 1996) which was presented in two volumes: *The Grey Slates of the South Pennines, Volume One: The Industry*, and *Volume Two: The Quarries and the Slates*. Volume Two includes the database of stone slate quarries and more detailed information than is reproduced here.[2]

PHASE II – Building the future

Phase II of the project was developed from the conclusions of the first phase of the work and was carried out during 1995. The scope of this part of the project is described on page 27. The conclusions and recommendations from these two phases of work are presented on page 28.

THE HISTORICAL CONTEXT: TRADITIONAL BUILDING MATERIALS IN DERBYSHIRE AND THEIR USE

Our knowledge of Derbyshire's buildings is based on the Statutory Lists of Buildings of Special Architectural or Historic Interest, archaeology and on the observations of casual travellers. Timber houses, similar to those described by William Harrison in the early seventeenth century, have been excavated on the site of the former village of Barton Blount. Deserted in the fifteenth century, the latest houses (dating from the fourteenth century) were primitive framed buildings resting on pad-stones and had a life of at least 50 years. A saw-pit containing oak sawdust was also discovered (Beresford 1975, 40).

Eighty years ago, in his classic work on English building construction, C F Innocent observed that: 'stone was only used for the foundations in ordinary English buildings, up to the time of the Renaissance, and walls of worked stone, or "ashlar", are scarcely to be found in the buildings of older England, except in those of importance ... There are probably few localities where the use of stone as the principal material for walls of minor buildings is of any antiquity' (Innocent 1916, 118–9).

This view of the impermanent and mean construction of the earliest vernacular houses is increasingly being confirmed by archaeology. 'Rural vernacular houses prior to the late Middle Ages appear, from the evidence of excavation, to have been of uniformly poor quality throughout the whole of England' (Mercer 1975, 8).

Certainly, no yeoman houses and few minor gentry dwellings (whether of timber or stone) have survived in recognisable form from the sixteenth century or earlier in the stone areas of Derbyshire. However, tree-ring dating is increasingly revealing fifteenth- and early sixteenth-century origins for the more substantial gentry houses. Further south in the county, there are significant remains of superior timber dwellings. Dating generally from the sixteenth century, they often survive as parts of later houses or, more widely, fragments of dwellings which were originally timber-framed and are now encased in later stone or brick shells.

Over 150 cruck-framed buildings also survive, distributed mainly over the east and south of the county, although by no means unknown in the west and north-west. They range from the simplest and smallest

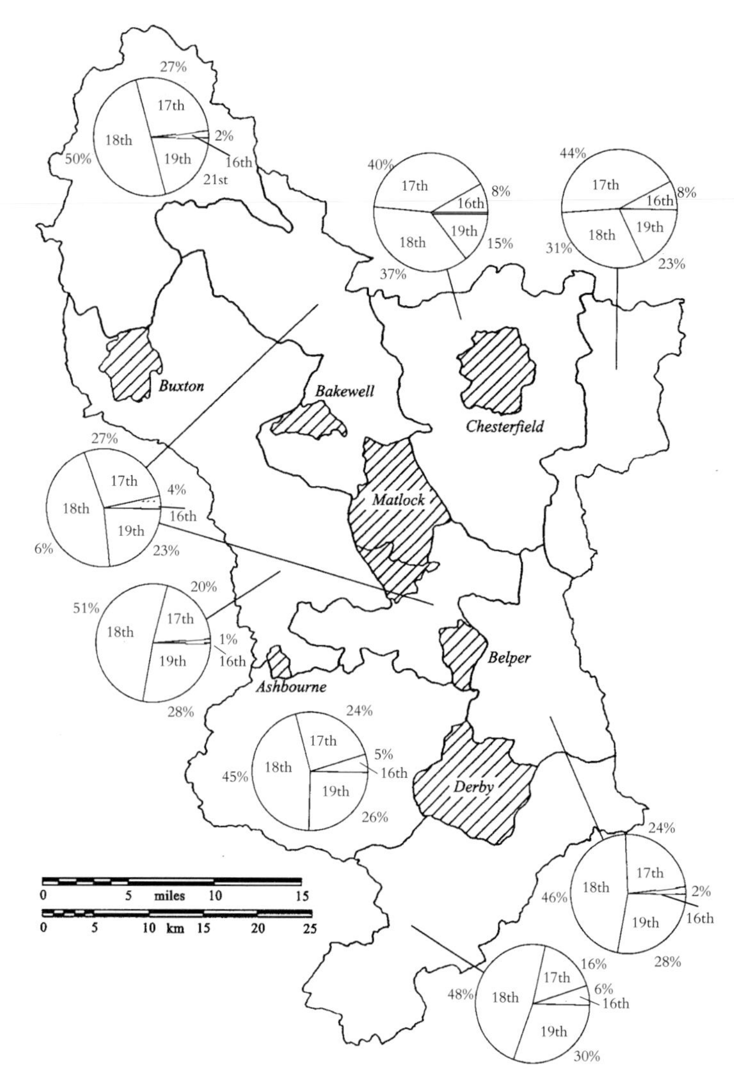

Figure 1. Derbyshire listed buildings: distribution of sixteenth-, seventeenth-, eighteenth- and nineteenth-century houses (© English Heritage, Derbyshire County Council and the Peak District National Park Authority).

surviving vernacular houses to the cruck barns which are among the county's most impressive buildings. Tree-ring dating is again demonstrating that many of the cruck barns date from the fifteenth and early sixteenth centuries. However, more fieldwork and research are needed to confirm to what extent the stone-region crucks were associated with timber walling or whether (as now) they were always seen in conjunction with walls of stone (Alcock 1981).

An early seventeenth-century traveller through the Peak might have noticed, here and there, brand new or significantly rebuilt farmhouses in stone, perhaps still with roofs in straw thatch, but more often in stone slate. Elsewhere, away from the stone districts and possibly some years later, a similar rebuilding would also take place, perhaps in timber but more likely in brick. Many historians have studied and documented this apparent 'Great Rebuilding' of (mainly rural) England, which began at various dates at the highest social levels and was driven by complex reasons. Recent research has demonstrated that this major rebuilding was taking place along a 'moving frontier of modernisation' and continued at least until the end of the eighteenth century (Platt 1994, 2). Its implications for the use and provision of building materials were considerable. By the end of the sixteenth century, timber was already in short supply over parts of the county and could only be replaced by local stone or locally fired bricks. The economic transport of quantities of materials from further afield was still nearly 200 years away.

The expansion in quarrying had almost certainly begun before the end of the sixteenth century, brought about by the rebuilding and remodelling of houses at supra-vernacular and gentry level. At Hardwick (Old Hall in the 1580s, New Hall in the 1590s) and Haddon (throughout the sixteenth century), new or major rebuilding in stone was taking place. Delves near Rowsley and on Stanton Moor provided stone for the alterations at Haddon; at Hardwick, stone came from a delph within the Park and stone slate for the Old Hall was supplied by delves in Walton, Whittington and elsewhere (Durant & Riden 1980, xxiv). In the south, Longford Hall (sixteenth century), Sudbury Hall (*c* 1620 and 1660s), and Weston on Trent Hall (1630s) were all of locally-fired brick. Bess of Hardwick's first house at Chatsworth (completed 1576) had been an ambitious edifice in brick while at Repton, Prior Overton's Tower (1436) represents the earliest brick house of the region (see also Craven & Stanley 1991).

The contrast between the old and new ways of building is nowhere better seen than at Hardwick. During September and October 1591, John Worthington was getting and dressing slates 'at the old Feld', 'Walton Spring', and 'Walton Hey' for the roof of the Old Hall. Yet some two years later, payment was being made 'to the plumer ... for coveryng the new buylding roof' (at the New Hall) in lead (Durant & Riden 1984, 210) While the parapeted roofs of the newly fashioned houses were in lead, for other, more traditional new or rebuilt dwellings the choice of roofing probably remained much the same as it had been for their predecessors. Only with the coming of the canals in the late eighteenth century and the railways some 50 or so years later were the options for roof covering widened to include slates, whether Swithland, Welsh, Cumbrian, or from even further afield, and tiles, whether Derbyshire, Staffordshire, or pantiles. The consequences of this widened choice are reflected in the character of the roofs we see today.

The character of Derbyshire buildings – evidence for a seventeenth-century rebuilding

The Derbyshire Lists include over 280 vernacular houses with inscribed dates within the period 1600–1800. Dates are found chiefly on door lintels but other dated features include cast rainwater goods, rectangular or oval plaques set into walls, plasterwork and glazed headers set into brick gables and walls. A date profile for all the county's vernacular buildings, plotted for its various regions, is shown in Figure 1; the chosen areas broadly reflect the major geological divisions. Those regions for which the lists have insufficient dating information are omitted.

The profile clearly demonstrates that the northern half of the Carboniferous and the Magnesian Limestone have the highest density of sixteenth- and seventeenth-century vernacular stone buildings, and these are largely stone slated. Presumably, these are the survivors of the rebuilding. Elsewhere, although earlier buildings survive, the picture is one of mainly eighteenth-century construction and rebuilding in brick with tiled roofs. While the economic climate must have influenced the ability of the rural population to rebuild or improve their own houses, other factors could also be involved. Northern parts of the county had dual economies, with industry creating wealth alongside agriculture. It is chiefly in these areas that farmers and the minor gentry were able, in the sixteenth and seventeenth centuries, to replace or perhaps significantly rebuild their homes in stone. Elsewhere in the county, although such rebuilding had also begun, it was not until the eighteenth century that new or replacement building in more permanent material, sometimes stone but mainly brick, began to take place.

The use of building and roofing materials

Evidence for an apparent expansion of stone delving in the sixteenth and seventeenth centuries is provided by the field names associated with that activity (Fig 2), demonstrating the extent to which quarried stone was available throughout the upper and lower Derwent Valley in particular. At what date delving could be considered an industry is uncertain since, in most instances, the taking of stone from the 'commons or wastes' for local use was enshrined together with other customary rights of the manor. Opening up delves for stone supplies (or the removal of other building materials from the commons or wastes) was subject to only the loosest of control through the manorial courts.

However, stone for sale outside the manor or to other tenants or freeholders did command a value. In 1388, a third part of the quarries of Lord de Vernon in Baslow was assigned for the dowry of Juliance Vernon (Kerry 1901, 4). In the manor court rolls there are also numerous references to the letting of the quarry. By the time of the Dissolution (if not significantly earlier), the demand for stone for monastic and church building must have been in serious decline. The county's few great feudal houses would represent only an intermittent demand and the problems of transport to any distance would almost certainly rule out the use of stone for any but the most local of building projects. The delving and use of stone slates too must have been subject to similar cycles and constraints.

In 1421, John Bullock of Norton (then in Derbyshire) was permitted to erect buildings within the manor and 'should have such timber as exists within the desmesne there ... [and] ... should have Sclatestones and any other stones within the desmesne' (indenture of lease, Sheffield City Library Collection, 651). Here is a suggestion that, in the fifteenth century, where stone slates were locally available they were used to roof the houses of the gentry

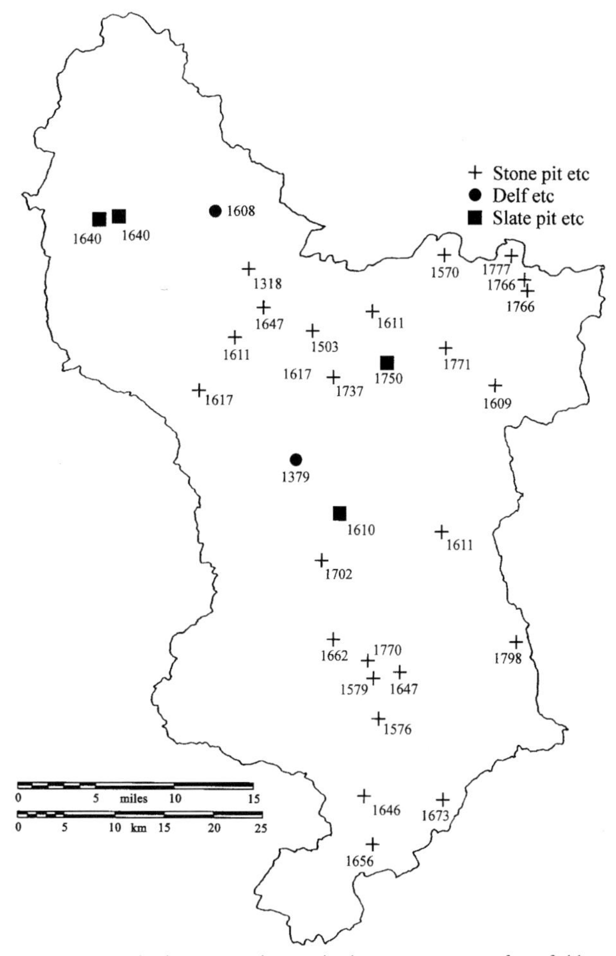

Figure 2. Derbyshire stone slate and other stone quarries from field-name evidence ((© English Heritage, Derbyshire County Council and the Peak District National Park Authority).

(and perhaps of others). We know nothing of the nature or size of these early slates. Archaeology has recovered 'four complete stone roof-tiles, and fragments of a further fifteen ... eight were Charnwood (Swithland) slate, five from the Magnesian Limestone, three from limestone, one from mudstone and one from sandstone' on the site of the South range at Dale Abbey (Drage 1990, 75).

Earlier excavations here (carried out by H M Colvin) had also produced stone-roofing slabs with holes as well as Swithland slates and ceramic roof and ridge tiles. Unfortunately, neither report gives any information on size. Elsewhere, stone slates in thin lamellar sandstone were found from field-walking (systematic examination of the surface soil) on the site of a possible fifteenth-century former house of the Okeovers (just in Staffordshire); however, they were too fragmented to size. Place-name evidence (Fig 2) offers only four specific references to 'slate pits' directly, but it is highly probable that some of those described from the sixteenth, seventeenth and eighteenth centuries might also be able to provide lamellar gritstone or sandstone suitable for slating. Farey's 1811 list of stone slate quarries includes some 13 which were also described as freestone quarries (Farey 1811, 1; and see Fig 3).

The extent to which a rebuilding in stone had gathered pace by the early eighteenth century in the mid and south parts of the county is perhaps suggested by William Woolley in his *History of Derbyshire*. In Staunton [Harold],

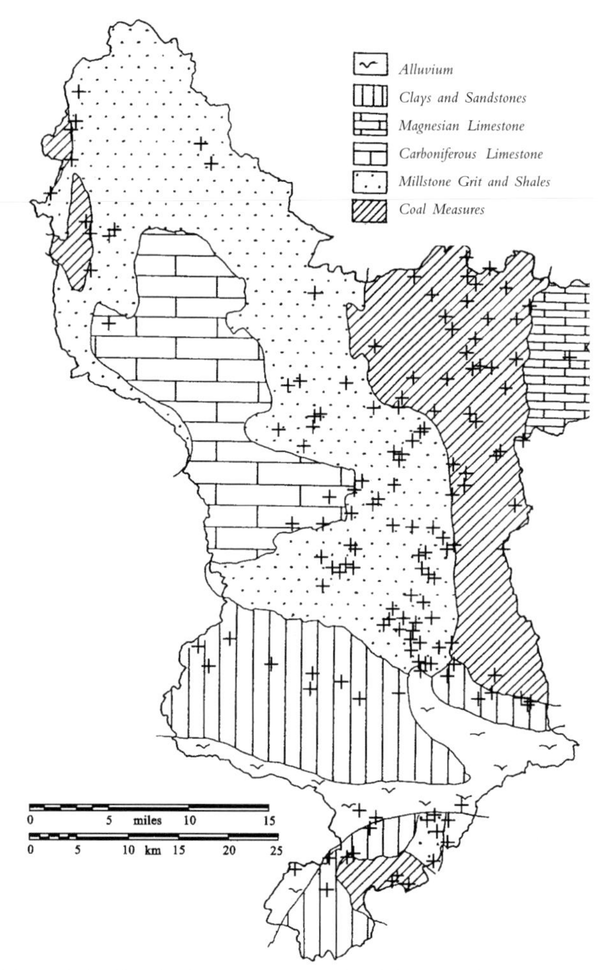

Figure 3. Derbyshire freestone quarries: listed by Farey (1811 © English Heritage, Derbyshire County Council and the Peak District National Park Authority).

he notes 'diverse stone quarries' while Hartshorne had 'quarries of freestone, limestone and coal' and, further north, Duffield Bank was described as 'a pretty large ridge of hills between Holbrook and Horseley full of stone quarries'. Breadsall had 'some quarries of gritstone which works well and is good for paving' (Glover & Riden 1981, 6).

Compiled a century later, Farey's list of freestone quarries (Fig 3) shows the extent to which quarrying for building stone had extended (Farey 1811, 416–22). All the geological measures were then being exploited and, even allowing for any deficiencies in his list, few areas of the county were not within a few miles or so of a building-stone quarry. Only with improvements in roads locally and the coming of canals and railways could dimensional stone be moved in significant quantities. Farey also supplies some of the most comprehensive comments on the variety of building materials in use over the county.

> Over about two-thirds of the County of Derby, Limestone or Grit-stone Buildings almost universally prevail; the Walls being very substantially built, and gable-ends, with chimnies in the end walls, being nearly general; in the Red Marl and Gravelly districts at the south end of the County, and in the Coal-district there, and in part of that near the eastern side of the County, Bricks are generally used in the walls; red, and of a pretty durable kind ... The Roofs of the Buildings in Derbyshire seem sharper pitched, or more acute at the Ridge, than is usual in the south of England, particularly, in perhaps one-third of their whole number, which are covered with the grey and white Slates or Tile-stones of the district ... The Red Tiles made in Derbyshire, perhaps for want of washing and more perfectly tempering their Clays are found less durable, as well as thought less handsome by many, than a sort of black or very dark blue dull glazed Tiles, almost the colour of new cast Iron, which are brought out of the Pottery district of Staffordshire ... and are pretty extensively used, in the southern part of the County.
>
> I saw but one remaining instance, of the Shingles of cleaved Oak, or wooden Tiles, which probably was once much more common, and that was on the Church at Walton on Trent: this method of covering Buildings, Churches in particular, still continues in Sussex, Kent, Essex, and some other Counties, to waste some of the very best of our Oak Timber, and ought to be speedily discontinued. No considerable quantity of the Straw of Derbyshire, is, fortunately, diverted from the more important purposes of litter and manure, to cover permanent Buildings; and the few Thatched Farm Houses and Cottages which are still found, ought to give place to Tiles, or Tile-stone coverings ... In the northern parts of the County it is common, when Thatch is used, to lay it on a course of strong Eaves-slates or Tile-stones, which prevent the Cattle from pulling the Thatch off low Buildings surrounding a yard, or against a field; and ladders, &c. occasionally place [3] against such Buildings, do less damage to thatch Eaves; and about Mansfield, Notts, such Eaves-slates are used to Tiled Buildings with the same view. (Farey 1811, 12–20)

By 1830, English annual production of plain clay tiles was in excess of 41,700,000, and that for pan and ridge tiles 20,600,000. Staffordshire production in 1832 was estimated at over 11 million (*Statement of Duties* 1830-1, 345). In the north-east and east of the county, the highly localised use of traditional pan tiles relates to their long-established use further east in Nottinghamshire and Lincolnshire. Locally, their use occasionally overlaps that of stone slates and in a few instances stone eaves slates are associated with pan tile roofs.

The distribution of these traditional roofing materials is shown in Figures 4, 5, 6 and 7 and summarised in Table 1, and confirms the general conclusions of writers such as Brunskill who have attempted to identify their use and distribution (Brunskill 1978a, 41–65; Brunskill 1978b, 194–6) The data also demonstrates the rate of their loss resulting from unauthorised roof repairs or substitution, or perhaps from reroofing before listing.

The surviving thatched buildings (Fig 4), mainly in the area south and west of Derby, may be divided into two distinct groups. Notwithstanding Farey's comments above, there remain a few older houses, cottages, and outbuildings (generally timber-framed and presumably always, or for

Table 1. Surviving traditional roof coverings in Derbyshire on listed buildings (see Figure 1 for omitted areas).

Pantile	Plain Tile	Slate	Stone Slate	Thatch	Modern & other
30	1075	927	770	33	168
1%	36%	31%	26%	1%	5%

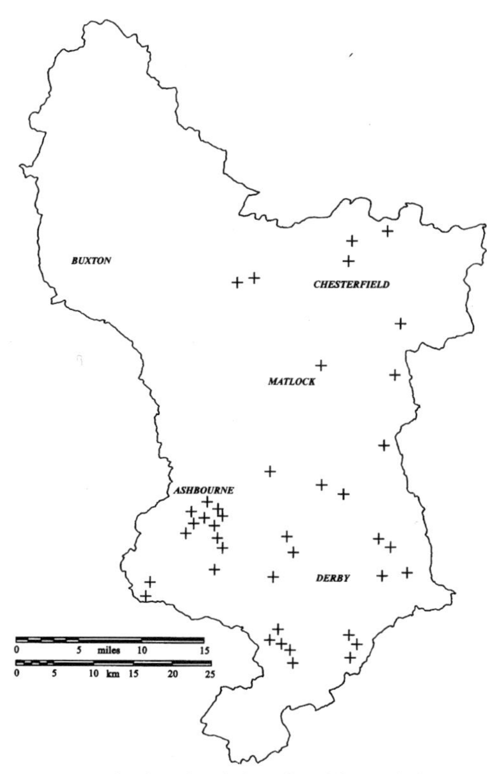

Figure 4. Derbyshire thatched roofing ((© English Heritage, Derbyshire County Council and the Peak District National Park Authority).

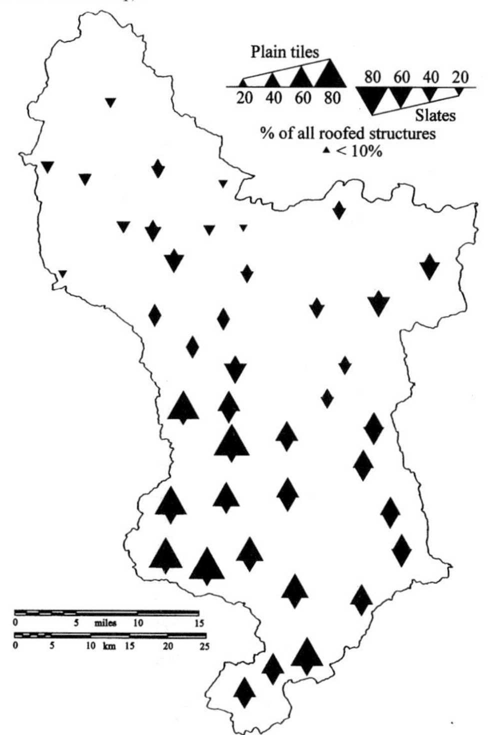

Figure 5. Derbyshire slate and plain tile roofing ((© English Heritage, Derbyshire County Council and the Peak District National Park Authority).

many years, thatched) and to the south of Ashbourne there is a group of mainly nineteenth-century thatched estate cottages in the vernacular tradition. Innocent (1916, chapter xiii) makes much of the traditional forms of thatching which still survived in the region in the early years of the twentieth century, but regular rethatching using alien materials and techniques has almost completely displaced them.

In Derbyshire, plain tiles and metamorphic slates are the roofing materials of the Industrial Revolution (Fig 5).

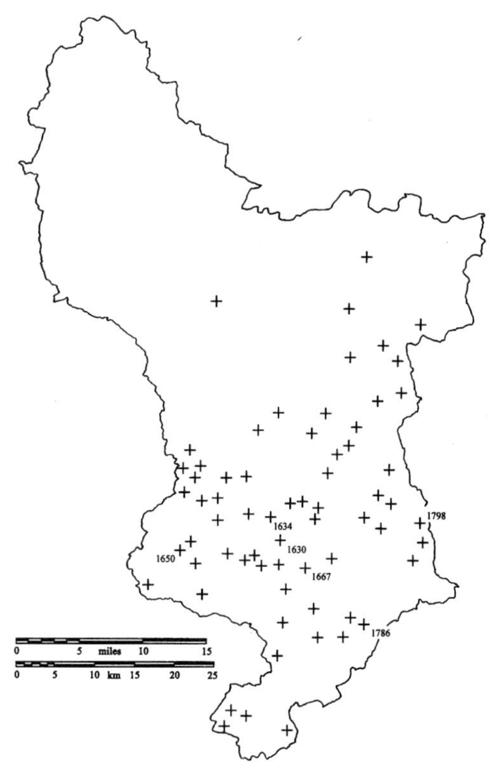

Figure 6. Derbyshire brick kilns ((© English Heritage, Derbyshire County Council and the Peak District National Park Authority).

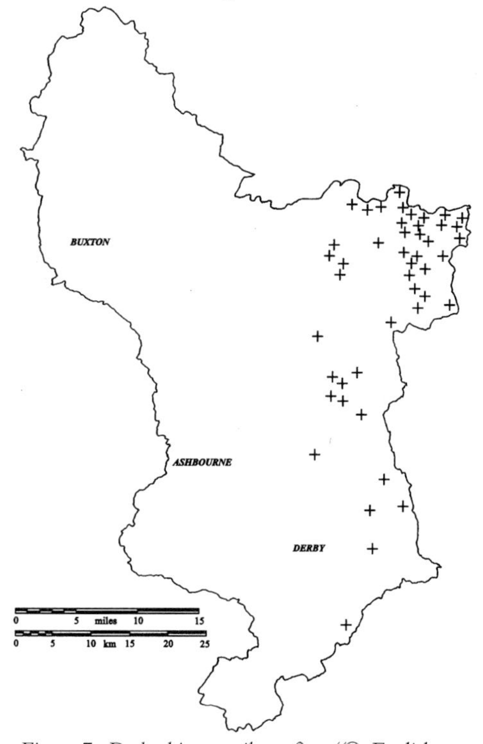

Figure 7. Derbyshire pantile roofing ((© English Heritage, Derbyshire County Council and the Peak District National Park Authority).

It may be possible that Swithland Slates (north Leicestershire) were being used in the south of the county before the late eighteenth century, but unfortunately the List descriptions do not differentiate between these and similar slates from elsewhere. Just a few roofs, particularly around Melbourne, are described as being of red plain tiles. Further research into local brick-makers engaged in late seventeenth- and early eighteenth-century building and rebuilding in south and south-west Derbyshire may

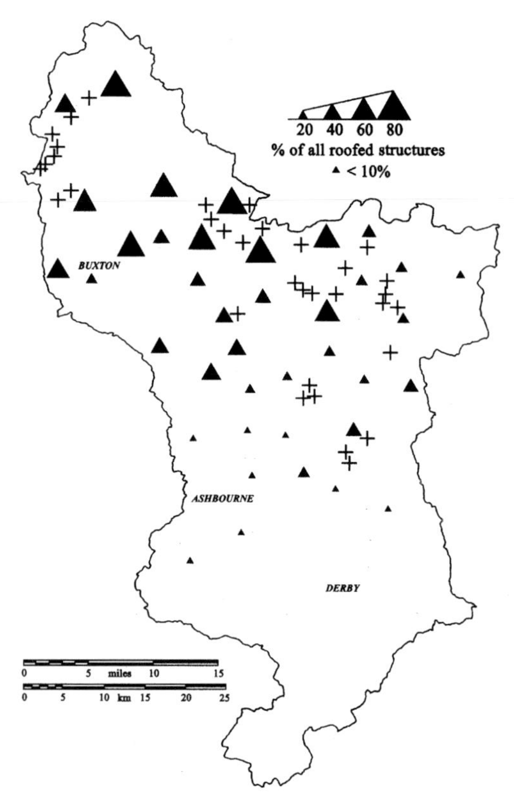

Figure 8. Derbyshire stone slate roofing ((© English Heritage, Derbyshire County Council and the Peak District National Park Authority).

reveal the extent to which they were also providing the red plain tiles. These were condemned by Farey before their replacement by tiles from Staffordshire. The extent of this activity is well illustrated by the survival of field names associated with brick making (Fig 6).

The use and distribution of plain tiles is confined to the southern half of the county in an area south of a line through Matlock. For parishes in the western half of this area, they represent almost the sole form of traditional roof covering (a little thatch excepted); in the parishes further east, approximately 50% of roofs are in plain tiles and 50% in slates. North of Matlock (and especially the north-west), when roofs are not stone slate they are likely to be of slate. In the north-east, no one material is dominant, with pantiles, slate and stone slates all surviving (Table 1, Figs 7 and 8).

Stone slate roofing

The distribution and density of use of stone slate roofing is shown in Figure 8. While the highest densities are to be found in the parishes of the north-west, the upper Derwent Valley and the 'non-pantile' parishes of north-east Derbyshire, their use extends fully over the northern half of the county. Outside the northern areas, where stone slate roofs form from 50 to 80% of all roof coverings, they are (in the more southern examples) generally at less than 10% density.

Figure 8 suggests that the areas of highest rates of survival (and presumably original use) of stone slate roofs are still, in general, closely associated with the areas which had the highest density of delves (Farey 1811, 428–30). If we accept Farey's list as comprehensive then, in the absence of nearby delves just over the boundary in Staffordshire, Figure 8 suggests that, in west Derbyshire at least, stone slates were being transported distances of up to about 15 miles. The missing data for the mid-Derwent valley arises from the deficiency of information on roof coverings in the Belper and Matlock old Urban District area Lists. The predominant stone slate roofing areas are also those where the highest density of modern replacement roof coverings are to be found.

Overall, for the areas included in the survey, approximately 5% of all listed roofed structures have their original roof coverings. Although we can never be sure what the original materials were in all cases, it seems that, with the highest losses in the predominantly stone slate areas, these are the roofs which have in the past been most at risk from replacement with modern materials.

THE STONE SLATE INDUSTRY TODAY

At the start of this survey in 1994, the industry within the study area had declined to the point of extinction. There was some production in Yorkshire and supplies of new stone slate for the South Pennines were largely dependent on the quarries in that area. Happily one quarry, MacStone at Kerridge, has since begun trial production and another, very small, delph has been discovered operating at Fulwood Booth, near Ringinglow. These delves exemplify the range of activity within the industry today. However, each can only supply one of the seven slate types which this study identified as occurring in the region.

It had also been assumed at the outset of the study that the manufacture of stone slates had been discontinued early in this century. In fact, there appears to have been a significant level of production until the 1950s, when the boom in house building, and the expansion of concrete-tile manufacture which it fuelled, almost wiped out the last remaining vestiges of the stone slate industry. However, competition from cheaper products was not a new phenomenon.

During the nineteenth century and the early part of the twentieth century, the Welsh and the Cumbrian slate industries had made inroads into the natural market for stone slates. The Welsh quarries in particular developed a product specifically designed to supplant stone slates: large and randomly sized in imitation of the local item, they were known as ton slates. Lighter in weight than the traditional product, they were not only cheaper to buy but also needed less timber to support them and could be transported on the developing canal and rail networks at acceptable cost. Later, they in turn were supplanted by the cheaper and even lighter single-size slates which became the speciality of the Welsh quarries. Their success is still evident on the mills and other industrial buildings of the region and on the houses which grew up around them.

Another difficulty for the manufacturers of stone slates is the longevity of their product. Even for roofs which need re-covering after 100 years or more, a large proportion of the original stone slates can be reused. So, unless there is a booming building industry using new stone slates, demand is small and whenever a stone slated building is reroofed with other products, the future market dwindles further, or the supply of second-hand slates is encouraged.

During the post-war period and until quite recently, reclaimed slates were also available in large quantities from the demolition of redundant factories. Although demolition has now declined substantially, it has not ceased altogether.

Another factor in the decline of the stone slate industry was the loss of skilled workers during the two world wars. Contemporary accounts describe the difficulties of maintaining production as the workforce moved to munitions factories and other essential industries. Even after 1945, the industry was complaining of the difficulty in recruiting staff to work in the arduous and uncomfortable conditions which prevailed in many quarries. With insufficient labour, production declined and shortage of supply was added to the reasons for choosing other products. Thus, competition from cheap products, the availability of second-hand stone slates and the loss of skilled labour all contributed to the decline of the industry. Fortunately, a small part survived and with it the skills necessary to bring about its regeneration.

The following sections deal primarily with the manufacture, marketing and sales of stone-roofing slates. However, it is argued elsewhere in this chapter that production of stone slates alone is not feasible, or may at least be uneconomic. So comment is made where appropriate on other products, such as flags and kerbs, which would use production technology, and sales and marketing methods which are similar to those for stone slate manufacture.

Manufacture

Stone slates in Derbyshire are produced through the delving of the fissile rock, splitting to thickness, dressing to size and holing to provide a fixing point. Unlike the frost-spilt slates such as the Stonesfields of the Cotswolds or the Collywestons of Northamptonshire, production is relatively straightforward and not reliant on weather conditions.

Delving

Before any stone can be quarried, the overburden (the vegetation, topsoil and other material lying on top of the useable stone) must be removed. The overburden may be only a few centimetres deep and is usually removed by hand or with small back-hoes or face shovels. Where the slate rock is close to the surface, care is essential and it is only where the overburden depth is significant that large contractor's plant would be appropriate.

Waste is generally carried to the tip in a face shovel or small dump truck. Temporary tips are sometimes created to be cleared with larger plant at infrequent intervals. As far as removal of the overburden and transportation of waste is concerned, there seems little need for more sophisticated plant.

Extraction for production is also generally low-tech even at its most mechanised. This is not a criticism of the method but an acknowledgement that stone slate needs careful extraction. Manufacturers strongly believe that, once fissile rock is uncovered, it should be extracted by hand using bars, chisels or jack hammers. In view of the essentially fragile nature of the slate and the overwhelming need to produce slabs as large as possible, this is certainly the preferable method.

Normally, the rock is extracted by taking advantage of the natural beds and joints. Where large blocks of rock occur, they have in the past been pulled out of the quarry face with winches. This was always done slowly to avoid jarring the rock. The option to drill and blast is completely inappropriate for slate production since it would certainly shatter the slabs. Sawing by diamond wire or circular blade to extract rock, although less destructive than blasting, has little to offer since it would cut across the jointing and result in smaller slabs.

Today, the rock is sometimes worked in benches a few metres high and blocks are lifted from the face with a face shovel. Although, as might be expected, this does tend to break some of the largest slabs, it has the benefit of higher productivity and, in the delves where it is used, still produces reasonable sizes up to about 1m (39 in) long. Nonetheless, the method is probably unsuitable for most delves. Thus it is difficult to identify where improvements in technology could assist in mechanising delving.

Conversion

Traditionally, this was entirely a hand process and specialised tools have been developed for each step in the process. It involves:

- splitting to the final thickness with a chisel
- trimming the edges 'square' and bevelling with a hammer
- holing with a spiked hammer.

While there are still adherents to the hand-made tradition, a certain level of mechanisation has appeared in recent years. This has not always been welcomed by traditionalists because it has changed the appearance of the product.

The most revolutionary innovation is sawing block to thickness. This is potentially unsatisfactory, especially in bedded sandstones, for at least three reasons. Firstly, it encourages the use of any sandstone, even if it is of very poor durability and entirely unsuited for roofing. Secondly, it is almost impossible to ensure that sawing is parallel to the bed. This means that beds which would normally be on the surface or entirely within the thickness of the slate will now 'outcrop' as bands across the surface. Whereas these beds might not have materially affected the durability of the slate when correctly manufactured, exposed bands, if susceptible to weathering, will deteriorate very rapidly and perhaps last no more than a few years. Thirdly, it produces a constant thickness and an entirely flat and bland surface with none of the character which is so important to the traditional product.

Sawing is also used to size slates. The claimed logic here is that it is cheaper to saw a block to size before splitting it to thickness. This is not necessarily so and its adoption may stem more from a familiarity with the advantages of sawing for masonry than with an understanding of slate manufacture. It has the drawback that it makes the bevelling of the edges an extra task at added cost and encourages both the manufacturer and purchaser to regard bevelling as undesirable, despite it being an important characteristic of the traditional product. Sawing also tends to reduce the range of widths available to the slater, making his task more difficult and the roof more uniform. Additionally, it denies the logic of maximizing production from the quarry. On balance, the cost of sawing and subsequent bevelling may be greater than for the traditional method.

In most cases, bevelling is still a hand process, although one manufacturer has developed a machine to carry out the task. This development is very desirable as it not only increases productivity but also creates an authentic edge. It would be of benefit to the industry generally to adopt this innovation.

Holing

Traditionally a slow and careful hand process, holing was carried out by specialists travelling from quarry to quarry. In most cases, the job is now left to the roofer and is generally done with an electric drill.

The traditional hole was hourglass-shaped in section, having been formed from both sides. This helped to grip the wooden peg from which the slate was hung (pegs were dried before use and as they absorbed moisture they swelled and became secure in the hole). With the advent of large nails driven into the battens, the tapered hole was no longer required. However, if slates are hung over the battens on metal pegs, then it is essential that the holes should provide a close fit to the peg. This will prevent it from tilting in the hole and allowing the slate to slip. Alternatively, the peg can be held in place with a second batten or by torching (provided under-slating felt is omitted). On balance, mechanically drilled holes are acceptable and are cheaper.

Mobile production

It is recognised that some slate types will probably only ever command a small market. The extreme example is the Whitwell type of the Magnesian Limestone of which there are only a few remaining roofs. Others might include the small stone slates of Abney or the pink slates from Wirksworth. The solution to producing such small volumes is intermittent delving using a mobile operation. This would typically comprise a small excavator, a builder's dump truck and a secure cabin in which to store tools and product. Such a system could be operated out of a permanent quarry and, with the addition of a more sophisticated extraction system (for example, track-mounted circular saws), might be equally appropriate for production of masonry stones or metamorphic slates with similar limited demand, such as those at Swithland in Leicestershire.

THE DELVES

During the earliest period of the stone slate industry, development would have been greatly influenced by the difficulties and cost of transporting a bulky and heavy product. Consequently, production would generally have been very localised. Wherever suitable rock was exposed, it would have been exploited, perhaps to supply a market no larger than the nearest village or even just one farmstead. As the industry developed and transport improved, some delves became large enterprises, especially those near areas of urban development.

The study included the preparation of a database of records for some 200 quarries in the region (Hughes 1996, vol 2). Farey's *A General View of the Agriculture and Minerals in Derbyshire* (1811) was the basis for this work, providing two lists of quarries which were producing slates and flagstones in the region in about 1800. Given that Farey's objective was to describe the scale of the industry, the references for the production sites were both useful in locating the quarries and, in retrospect, provided an impressively complete picture of the extent of stone slate manufacture at that time. In 1993, Michael Stanley carried out a review of Farey's list, giving specific map references for many of the latter's sources; this proved invaluable, not only in reducing the time spent searching for quarries but also in showing an overview of their distribution in relation to the geology of the region (Stanley 1993).

The third source of information on the location of fissile rock was geological studies. The region has a long history of these, with publications in the form of geological memoirs and maps (see Hughes 1996, vol 2, annex 2.9). A formal review of this literature was carried out for the associated study (Hughes *et al* 1995).

Among other useful sources of information were Ordinance Survey (OS) maps at 1:50 000, 1:25 000 and 1:10 000 scales, and facsimile reproductions of the first edition OS maps in the region (1842). For a general overview of the region and for mapping of locations, the OS Touring Map of the Peak District covered most of the study area. As part of the environmental impact study, 1:25 000 moor and heathland maps were consulted. Useful as the maps were, the variety of scales employed (especially in the geological maps) was a cause of considerable difficulty and frustration.

Besides indicating the specific delph sites, the maps also give place names such as Slatefield and Slatepit Dale which imply the presence of old delves. Unfortunately, place names persist longer than the activities they describe and many presumed sites have now disappeared without trace. This is particularly the case in the Coal Measures in the east of the region, where open-cast coal mining and the subsequent reinstatement have sometimes removed not only the quarry but the whole sandstone stratum.

The earliest written references which imply stone slate operations in the area, or slaters as a specific occupation, relate to the mid-seventeenth century and are in the form of agreements permitting the taking of slate and stone for the repair of buildings (Norfolk Estate papers: lease 1.7.1658).

A particular feature of stone slate delving was the tendency for operations to exploit relatively thin near-surface deposits. Two types of operation are apparent in the old delves. The optimum site was the crest of a scarp slope where the dip of the rock was parallel with the land surface. This allowed the delph to be developed into the dip slope without excessive overburden, sometimes even working two or more galleries, but mainly extending along the crest perpendicular to the dip. The classic example is Cracken Edge.

Where development along a crest was impossible, an open-cast system was followed. An extreme example can be seen at Harden Clough, where excavations extend over about 1 km^2 (0.386 square miles) as a series of shallow delves along the *c* 390m contour. In practice, the outcome often shows little sign of planned working, as at Freebirch. This system seems to have only been operated where there was negligible overburden.

Most workings produced large amounts of reject material. This was then dumped down-slope where possible, or into worked-out areas with virtually no attempt being made at restoration. In the latter case, a complex of highly disorganised excavations and dumps developed, rarely exceeding 7 m (7.5 yds) in depth below the surface or height of the tips. In such delves, a consequence of this system is that remarkably few working faces are now visible. Typical examples are sites at Wet Withens, Harden Clough, Thornseats, Brown Edge and Freebirch. Another explanation for the disorganised tipping on some sites comes from early records, where there were complaints by landowners that the managers, who were paid by the area worked, were tipping waste onto virgin land so that they could claim it as worked ground.

Virtually all the extraction processes were by hand, with the only concession to mechanisation appearing to have been in the form of narrow-gauge rail systems (operated by men or horses) at some of the larger workings. These systems were used mainly for the removal of waste rock, some of which appeared to be good building stone. Cracken Edge and Slatepit Moor Quarry, near Mossley, could also claim the advantage of rope-hauled inclines. Some of the larger sites had extensive areas of cutting sheds, together with barracks (worker accommodation) and, probably, a smithy.

The frequency and/or distribution of workings within a given horizon (a layer of soil or rock with particular characteristics) was influenced by a combination of the quality of the raw material, commercial factors (proximity to markets, competing materials, local practice) and the extent of accessible outcrop or 'competing' outcrops.

The reasons for an operation's closure may be pertinent. Excessive overburden, ownership boundaries and the introduction of competing materials (notably Welsh slate or Staffordshire tiles) were the main factors. In a few cases, when overburden became too great quarries extended underground as mines, for example at Cracken Edge, Longnor, and possibly at Spout House Hill.

Fieldwork

The objectives of the fieldwork were to:

- locate the quarries
- establish their local geological context
- confirm the presence of fissile rock
- obtain samples of slate rock to establish the typical features: colour, visible minerals, thickness, and texture (including grain size and features such as ripple marks)
- assess the potential constraints on production: overburden and the dip of the beds, access, historical interest and environmental and social impact (including visibility and proximity to housing).

The searches were organised around the list of delves obtained from earlier studies, but any indications of previous workings were followed up as they were encountered. Indicators included disturbed ground (either tips, excavations or old tracks) and isolated plantations or copses often resulting from natural regeneration. It is surprising how sensitive the eye becomes to such 'alien' features in the countryside, even when travelling by car or scanning a wide area with binoculars. Because many old workings are overgrown, it is easier to carry out such surveys when trees are bare of leaves.

Assessment of the delves was entirely qualitative and based on the presence of thinly fissile rock. About 190 delves were visited (not all slate producers) and neither time nor other resources allowed for quantitative work. Similarly, assessment of the rock's production potential, in terms of fissility and volume, was based entirely on the exposed faces. Consequently, the information in the quarry records should not be used as a basis for commercial activity without further detailed research and explorations.

At the outset, it was intended to include information regarding the ownership of the land or mineral rights but it soon became apparent that this, too, lay beyond the available resources. For these and other reasons, the presence of a delph in the database or in the individual records does not imply in any way that permission for quarrying has been, or might be, granted by the owner of the land or of the mineral rights, or that it would receive the approval of the planning authorities.

Colour photographs were taken of all significant sites (see Hughes 1996) and of examples of slate roofs in the immediate vicinity of the quarries. A set of photographs of a selection of the slate samples were also taken in diffuse daylight and included appropriate Munsell Soil Colour Charts (Munsell 1992). The geological report contains thin-section and back-scattered scanning electron photomicrographs. A library of sample slates and rocks created from this research is held at the National Stone Centre, Wirksworth.

The database

The field records were transcribed onto individual record sheets and the geological impact and other data records added. The database includes information on:

- reference number
- quarry name
- locality
- sub area
- six-figure grid reference
- geological horizon
- sample stone slate reference
- whether listed by Farey
- presence of stone slate

- presence of flagstone
- rock colour.

The region's sandstones which have produced roofing slates extend in a broad sweep from the Roaches in the south-west, around the moors of the northern Peak Park, and down through the eastern Coal Measures to the region south of Matlock.

The database contains 167 quarry records ranging in size from excavations no larger than a few square metres, for example, on Shatton Moor (SK 188802), to 1 km (0.62 miles) of almost continuous quarry face at Cracken Edge near Chinley (SK 037835). It is clear from the remaining evidence that the sources listed by Farey were often simply locations rather than specific delves and were not necessarily substantial, either when he noted them in the early nineteenth century or subsequently.

Analysis of the frequency, distribution and size of the old delves indicates that the scale of production was mainly influenced by proximity to substantial markets. Thus the sparsely populated areas of the northern Black Peak contain few and mainly small delves, while close to the markets at the edges of this region there were some very extensive operations. Outstanding among these are Harden Clough (SE 145040) serving Holmfirth, Glossop Low (SK 059963) and Cracken Edge (SK 037835). As transport in the region developed, it is probable that the two latter delves would have supplied the expanding urban areas to the west from Macclesfield to Stalybridge.

A similarly substantial and extensive industry existed in the east of the region, serving Sheffield and Chesterfield, with some large delves taking advantage of the comparatively easily worked rock. A small one still operates on a minor scale. Elsewhere, delves were quite limited in size, although many would have been big enough to supply the present-day market of the whole region.

Limestone is not a common source of slates in the region, although some fissile stones are known to exist in the Turnditch area (SK 299470). East of Chesterfield, there was a roofing industry based on the Magnesian Limestone. This apparently centred around Whitwell (SK 5576), but there is now virtually no field evidence available of its scale or extent. Indeed, there appear to be only two buildings remaining with roofs of this stone. Two old delves were discovered, identified in the database as Q40 and Q41, but with so few buildings (probably only two) in need of such slates, the only prospect for re-establishing production would be as a single project to provide a stock to be used as required.[4]

The slates of the South Pennines

Visual characteristics

The geological report explains the origins of the region's sandstones. Their appearance (and perhaps also their durability) varied according to the conditions in which the deposits from which they were formed were laid down. On the other hand, there was little variation in the mineralogy of the source rocks (which formed the sediments which in turn became the stone slate). It has therefore proved impossible to define individual stones mineralogically.

For the slates subset of the database of quarries and rocks, the following characteristics have been recorded:

- grain size: very fine, fine, medium and coarse
- colour: white, yellow, buff, brown, pink, olive and grey
- surface features: smooth, rippled, bedded and biogenerated features
- minerals: visually significant mica, presence of carbonate or organic layers.

Size range and mix

Local evidence suggests that some delves would have been capable of producing larger slates than others and that this produces a 'fingerprint' on the roof. While this is certainly true in some cases and in a general way, more work is required to prove whether it is a reliable diagnostic factor in determining sources within this region. It is possible that the natural mix from a quarry may have varied with time and may have been modified by 'market' factors. For example, if it was necessary to transport a consignment over a significant distance, say more than one day's journey, then the inclination would be to carry the larger or thinner slates as they cover a larger roof area for a given weight. Conversely, small sizes require more labour to construct a roof and would therefore tend to be

Table 2. Generic types of stone slates in the South Pennines (presented in order of prevalence from most to the least common).

Generic type	Description	Locality in the region
Cracken Edge (Figure 9)	Textured, with or without stepped bedding, fine to coarse grained, white and buff to dark brown.	The north-west from Roaches to Saddleworth +
Teggs Nose (Figure 10)	Textured, with or without stepped bedding, fine to medium grained, pink.	The north-west from Roaches to Bollington +
Kerridge* (Figure 11)	Flat, featureless, without stepped bedding, fine grained, grey mica surface.	The north-west from Macclesfield to Bollington +
Yorkstone (Figure 12)	Flat, featureless, without substantial stepped bedding, fine to medium grained, buff to dark brown	The north and north east of the Peak Park, east Derbyshire
Freebirch (Figure 13)	Strongly textured or ripple-marked, fine to medium grained, buff to dark brown and olive to grey.	The east around Holymoorside to Unthank
Wirksworth (Figure 14)	Strongly textured, fine to medium grained, pink to red.	Matlock, Cromford, Wirksworth and Belper
Magnesian limestone (Figure 15)	Strongly textured or ripple-marked, fine grained, grey or pink, magnesian limestone.	The east in the area around Whitwell

* The prevalence of Kerridge type may be underestimated because the mica surface may disappear soon after installation as a result of weathering, rendering them equivalent of the Cracken Edge type.

\+ These types extend into the large conurbations of the East Cheshire plain.

Figure 9. Generic stone slates of the South Pennines: Cracken Edge type (© Terry Hughes).

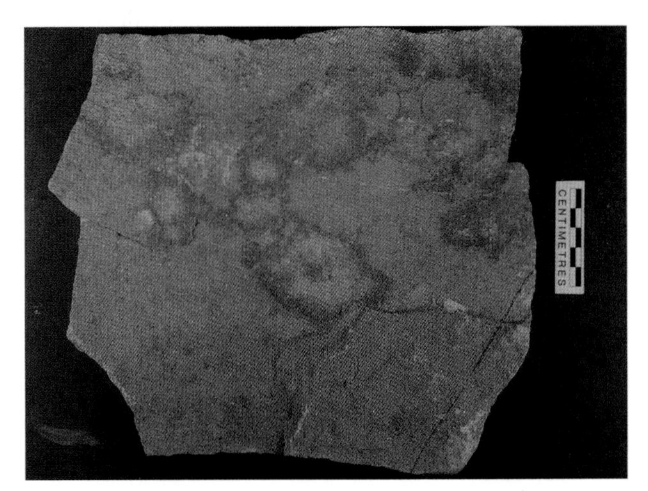

Figure 10. Generic stone slates of the South Pennines: Teggs Nose type (© Terry Hughes).

Figure 11. Generic stone slates of the South Pennines: Kerridge type (© Terry Hughes).

Figure 12. Generic stone slates of the South Pennines: Yorkstone type (© Terry Hughes).

Figure 13. Generic stone slates of the South Pennines: Freebirch type (© Terry Hughes).

Figure 14. Generic stone slates of the South Pennines: Wirksworth type (© Terry Hughes).

Figure 15. Generic stone slates of the South Pennines: Magnesian limestone type (© Terry Hughes).

used by more wealthy purchasers or by those for whom labour cost was less significant. Architectural conservation principles require the retention as far as possible of the original size mix when reroofing.

Constraints on slate types

Because depositional environments varied over quite short distances and altered in time, the grain size and shape and the structure of sandstones can alter significantly within a single small quarry. Therefore, it is quite possible that an individual site would have produced slates which varied in these aspects from time to time. It follows that it may not necessarily be possible to predict the continued production, nor to insist on the supply, of products with particular characteristics, except at the coarsest level of discrimination. Similar restrictions apply to colour, which can vary significantly within short distances within a single slate bed.

Generic slates

The slate characteristics above were analysed and from this it has been possible to describe generic types. Surface features are considered to be the most important characteristic, followed by colour, and grain size in so far as it determines texture. Superficial minerals may influence the appearance of the slates when new, but the processes of weathering and particulate pollution will rapidly modify this feature as well as the inherent colour. Size and the range of sizes are not included as generic features, although future work may show that these should be added to the descriptions.

Although the visible characteristics are of primary importance, less immediately apparent aspects such as mineralogy, grain size and chemical composition can also be relevant in determining the weathered appearance (including which species of mosses, lichens and so on establish themselves most readily on the surface). Durability is also very significant.

The generic types shown in Table 2 and illustrated in Figures 9–15 are proposed as a reasonable selection to cover the whole regional variety. Although they are named geographically, it is important to realise that they are a consequence of the depositional environment at the time they were laid down and therefore not necessarily restricted to that area. Indeed, some may be widely distributed within the region, the Cracken Edge type being by far the most common in the study area. This is one reason why slates imported from Yorkshire are not the most suitable substitutes for the traditional types.

Inevitably these types are a compromise: it is always possible to further subdivide. However, it is considered that they represent a reasonable basis on which to promote production of slates within the region. It is recognised that, in particular cases and for particular buildings, authenticity of source may be the prime consideration. In such instances, the library of sample stones could be consulted for the nearest match within the immediate locality. It is almost inevitable that any building being roofed for the first time will have been supplied from sources within a few miles. It is only in recent years that both the lack of local supplies and the availability of cheap transport have resulted in carriage over significant distances.

Potential sites for production

The delves listed in the database contain fissile rock (although they may not have records of slate manufacture) and form the basic list for assessment of the potential for re-establishing production. However, it must be remembered that many delves have been excluded from this list simply because fissile rock was not visible.

The research also indicated that owners/managers of existing active delves were regrettably largely unable or unwilling to produce roofing slates at present. Nonetheless it is recommended that intending manufacturers should look to these delves first as potential sources.

There are various reasons for removing sites from a list of those in which production might be encouraged. These reasons include:

- their presence in an area of historical or environmental significance
- close proximity to housing
- established use for other purposes such as recreational activities
- contamination by pollutants or dangerous substances
- difficult or dangerous access
- limited reserves or unfavourable topography.

Within each of these categories there may be a wide scale of impacts or constraints.

- The presence of a small craft industry within a Site of Special Scientific Interest covering many square miles could be considered to have an acceptable impact when balanced against the objective of protecting the vernacular architecture of the region.
- The extent of adjacent housing may vary from a single farm 1 km (0.62 miles) distant to a complete village or suburb which has grown up to surround the delph since it closed.
- Recreational uses may range from a seldom-used footpath skirting an old delph to a popular country park extensively reinstated with grant support and including an interpretative centre, guided walks and organised climbing courses on the quarry faces.
- Access may be capable of improvement and the impact of traffic for the scale of operations envisaged will be quite small, perhaps fewer than ten journeys a day by car and just a few a week for modestly sized commercial vehicles.

Consequently, each potential site must be assessed individually on the understanding that the inclusion of a delph in the database or in the individual records does not imply in any way that permission for delving has been, or might be, granted by the owner of the mineral rights or of the land, or that it would be approved by the planning authorities.

THE MARKET FOR STONE SLATES

Market size

All attempts to obtain a reliable estimate of the demand for stone slates have been frustrated by lack of information. Enquiries were made to local authority conservation departments within the region and to architects, merchants and roofing contractors in the area and in adjacent counties.

There are about 1000 listed buildings in the Peak Park and Derbyshire with stone slate roofs (Fig 8). [5] It proved impossible to determine how many of these had had their roofs renewed recently enough to place them outside the market for new slates. Within the important adjacent area of Macclesfield, which historically drew stone slate supplies from within the Peak Park, there are about 2000 listed buildings. About 750 of these have stone roofs, of which about 50% have been renewed, which leaves a market of something over 350 buildings. This figure does not include the buildings within the Conservation Area Partnership around Rainow, or Kerridge, which receives grant aid from Macclesfield Borough Council and English Heritage.

The information available from conservation officers varied considerably. They have records of listed or other designated buildings within their area but, where the authority does not operate a policy of grant aid, the records of other buildings (especially agricultural buildings with stone slate roofs) may be incomplete or non-existent. Sometimes the individual conservation officer's personal knowledge can fill the gaps. In areas where grant aid is available, there are records of reroofing which received such support but these were not always complete.

In Derbyshire outside the Peak Park, 25 reroofing schemes were completed between 1979–95 with grant aid totalling £165,000. Some authorities provide no grant aid but have achieved the conservation of at least some of the buildings within their area by ensuring action by the owners of properties included in the Buildings at Risk register.

Architects could not provide any useful evidence about the level of demand for stone slates. Although all those who had been involved with stone roofing agreed that it was difficult to obtain supplies of slates, generally they worked with so few stone roofs that their opinions could not be quantified. Data obtained from local architects is also an unreliable guide to the level of reroofing within the region, either because an architect is not part of the process or because those who are involved are not based locally.

A questionnaire was circulated to roofing contractors within Derbyshire and the surrounding counties, the recipients selected from the *Yellow Pages* and the membership of the National Federation of Roofing Contractors (NFRC). From *Yellow Pages* entries, those which were obviously not slaters were excluded and, from the NFRC members, only those listed as slating and tiling specialists were included. Of 114 questionnaires sent out, 24 were returned. It was carefully explained that the enquiry was about stone slates but, in spite of this, there was some confusion in the replies between stone and metamorphic slates.

As part of the questionnaire, the slating companies were asked how much stone slate they needed to buy from any source to complete an average year's work. The total for the 24 responding companies which carry out stone roofing was 2000 tonnes (1.96 tons) and accounted for a total of 114 buildings of all sizes. This is an average of less than 18 tonnes (17.71 tons) per building equating to about 123 m^2 (147 square yards) of roof. This estimate includes repairs, new roofs and extensions to existing properties.

The other important factor for the size of the market is that the stock of roofs which are currently being reconstructed or repaired can be assumed to be good for the next 50 to 100 years. Therefore, the market might be assumed to be dwindling away. In fact, this is not the whole story. A very substantial part of the stock of stone slate roofs disappeared during the 1950s and 60s when, because stone slates were unavailable, they were replaced with other roof products, predominately concrete interlocking tiles. These tiles have proved to be a very poor substitute. Not only are they very flat and regularly sized, they have also developed a covering of black moss entirely different in appearance to the lichens which grow on stone slates. Fifty years later, these roofs are coming to the end of their lives. If all the roofs in conservation areas were recovered with stone slates as they became due for renewal it would:

- provide a significant enhancement of a town or village
- reverse the trend of decreasing demand leading to declining supply
- provide the basis on which to sustain a stone slate industry.

Such reinstatement should be encouraged by being eligible for grant aid.

Supporting the market

Unfortunately, the conservation requirement to repair with authentic materials (supported by grant aid) has encouraged the deliberate removal of stone slates from existing roofs for reuse elsewhere, to the detriment of the regional roofscape. The solution to these problems is two-fold: improve the supply of new slates, and control the use of second-hand ones in order to eliminate their supply from unsatisfactory sources.

The three aspects of improved supply are sufficient production, control of stocks and price support. The first is dealt with elsewhere in this report; the second and third are interrelated and influenced by grants.

It may be confidently anticipated that any new slates will continue to face competition from second-hand materials, some of which will always come into the market legitimately and find a purchaser even if their prices have been reduced. However, unless the price of new slates drops so low that their production is not economically feasible, the price of reclaimed slates will remain high enough to perpetuate roof demolition or even theft. Market forces are therefore unlikely to prevent the continuing deliberate cannibalisation of roofs. Under these circumstances, it becomes necessary to impose some form of considered control to counter such cannibalization. Since it is almost impossible to regulate (or even identify) the source of reclaimed slates which are to be used on a building, the only option is to preclude the use of any second-hand ones except those reclaimed from the same building or group of buildings.

This would require careful supervision by the grant-giving authorities but might well be no more onerous than that currently required for other conservation and building-control purposes.

Another means of tackling this problem may be to pay the grant support directly to the slate supplier rather than to the building owners (as is the case now). The grant for the construction of the roof could continue to be made to the building owner. This would have the effect of making the grant money unavailable to second-hand sources while at the same time rendering the new product equally cheap to purchasers. There are options as to how grant money could be applied to achieve this and it could probably be made available almost anywhere in the supply chain. The most suitable options appear to be payments direct to the manufacturer or to an intermediary organisation acting as a 'wholesaler'.

Payments to suppliers could be made on the basis of material for specific buildings. The level of payment should preferably be based on the area of roof covered, as this is simpler to check than weight, or area of slate supplied. It would also avoid overpayments due to oversupply or underestimates of coverage.

At present, the supply of slates for roofing and reroofing is only loosely structured. New slates may be purchased by the building owner, general builder or the roofing contractor, and from a builder's merchant or specialist roofing merchant, or directly from the manufacturer. In such a complex structure, it may well prove difficult to control the payment of grants, a consequence of which could be high administrative costs. Therefore, it is concluded that, at least in the initial stages of the regeneration of the industry, an independent organisation might be the best choice to take on this role.

Such a system would help to eliminate some of the problems with the present situation. However, it might not achieve the level of confidence required to encourage entrepreneurs to start up production, or for existing manufacturers to create stocks. These objectives could be achieved by the creation of a central stock-holder. It is envisaged that a scheme could be created whereby the stock-holder would guarantee the purchase of each member manufacturer's production up to an agreed level and at an agreed price. Grant money could be used to fund the stock purchase and wholesaling activity but would be primarily applied to reduce the retail price. If grant aid was conditional on purchasing slates from the central stock-holder, it would avoid the risk of overproduction and hence competition from the member manufacturers.

It is acknowledged that such an organisation could be seen as a direct competitor to commercial businesses which currently sell wholesale and retail roofing and stone products. In so far as public money was being applied to the cost of the operation, this competition would be perceived as unfair. Three factors might mitigate against this: such an organisation might only be required for a short period until supply and demand become balanced, it could eventually become self-funding and the competitive commercial companies might benefit from its existence.

One means of achieving this would be to approach a suitable (possibly charitable) organisation to consider taking on the stockholder's role. Alternatively, a suitable body could be created under the control of the local authorities who currently administer the grant funding. The National Stone Centre at Wirksworth might be suitable and willing to undertake this role.

Marketing

Position

Within the roofing market, stone slates are a high quality product in terms of both aesthetics and durability. Ownership of a house with a genuine stone slate roof is regarded as desirable. This image is enhanced by the support conservationists provide for stone slate use and by the cost and rarity of the material. Since there is little prospect of the cost being substantially reduced, this image should be fostered by the manufacturers. To counter the product's initial expense, they should aim to position it in the top end of the market, emphasising its beauty, historical and conservation importance, durability and long-term economy.

Price

The product is expensive, both to purchase and install. Typical prices in early 1996 were £250 per tonne for second-hand stone slates and about £500 for new. One tonne is sufficient to cover about 7 m^2 (8.37 square yards) of roof on average, although there can be considerable variation depending primarily on thickness but also on size. Therefore, at present the cost could not be substantially reduced by the application of technology or by productivity gains. Consequently, it is unlikely that the point will be reached when grant support could be removed without a fall in sales. Nonetheless, there is a constant demand and roofers continually complain about the difficulty of obtaining supplies. Therefore, the present subsidised cost is not a barrier to the current level of trade and it may be confidently anticipated that these prices will not inhibit increased sales should the supply improve.

There is an important lesson to be learned here from the metamorphic (blue) slate industry: supply creates demand. At the end of the 1970s, the British roofing-slate industry was declining because of high prices and poor demand. But in the early 1980s, demand improved and this was supported by both increased home production and imports. Had the industry not responded in this way, the market would have probably continued to decline. In fact, demand continued to outgrow supply until the start of the recession in 1989. Since then, this market has recovered faster than any other roofing product (Fig 16).

The reputation for being expensive is unfair when the initial cost is amortised over the life of the stone slates. A study carried out in 1995 produced comparative lifetime

Table 3. Comparative life-cycle costs for roofing products.

Product	Maintenance period in years	Initial cost for supply and fix/m²	Repair & maintenance % of initial cost	Cost for 100 year cycle £/m²
Aluminium sheet	40	£32.50	2%	£102
Copper sheet	65	£42.50	1%	£73
Lead sheet	100	£55.00	1%	£59
Stainless steel sheet	100	£42.50	1%	£46
Zinc sheet	40	£37.50	2%	£114
Clay tiles	40	£33.00	10%	£113
Concrete tiles	30	£12.50	10%	£85
Fibre cement slates	30	£24.50	12%	£134
Resin slates	30	£28.00	12%	£149
Welsh slates	100	£46.00	12%	£56
Stone slates	100	£110.00	12%	£133

Sources: Comparison chart from F Moal (1995), 11 Roofing Products, Eurocom Enterprises Ltd and Davis Langdon Everest (July 1995), Industry Costs Update, AJ Focus

costings for a range of roof coverings. These are given in Table 3 together with those for stone slates. Taking into account the fact that stone slates will be reused at least once and perhaps several times, the lifetime cost is very much less than that over 100 years.

Price and coverage

Stone slates, in common with other random slates, are normally sold by weight with an indication of the area which can typically be covered by a given weight (assuming a given head lap, normally 75 mm [3 in]). To anyone unfamiliar with the system this is a confusing and cumbersome method of deciding how much slate they need. It is also a long-standing problem. In 1695, the anonymous author of *The Art of Building or an Introduction to Young Surveyors in Common Structures* (reprinted in 1945) complained that their proportions in 'covering houfes are uncertain, so I cannot affign a certain number, unlefs the builders fhall prefcribe a true proportion for the Slate' (anon 1659 [1945], part 4, 19).

It is difficult to be precise about coverage because there are many imponderables. Firstly, the thickness of stone slates varies considerably. Then there is the mismatch between the number of slates of each length in a consignment and the width of the building. It is the roofer's responsibility to make the mix of slates fit the shape of the roof, but inevitably there will be too many of some lengths to complete a whole number of courses exactly, with consequent waste and reduction of coverage. Further, for technical reasons, the head lap must be altered whenever there is a change of slate length, but the extent by which this reduces the coverage cannot be estimated without knowing how many courses there will be of each length.

The confusion surrounding coverage is made worse by the use of a mixture of metric and imperial units: weights are given in tons or tonnes, areas covered in square metres, square yards or squares (100sq ft). Still worse, suppliers are tempted to over-estimate the coverage of their products in an effort to appear cheaper. At best, the upshot is that the building owner is confused, frustrated and unable to make an objective comparison of costs. At worst, the purchaser and supplier end up in dispute about shortfalls in promised coverage.

In marketing terms, this is typical of a production-orientated industry: 'It's the customer's problem', 'It's always been this way', 'Why should we care?'. Units could easily be standardised: the construction industry has been metricated long enough for everyone to be familiar with metres and tonnes. Considering the small volumes involved, it would also be perfectly possible for a manufacturer to list the total width of each length of slate in a consignment. This has long been the practice in the American slate industry. The roofer will have to do this anyway, so any manufacturers providing such information will be improving their competitiveness.

Of course, a really competitive manufacturer would give serious consideration to supplying sufficient slate guaranteed to cover a given area on the basis of a constant 75 mm [3 in] head lap. The arithmetic, though a little complicated, is well within the capabilities of a simple computer spreadsheet. This would remove most of the uncertainty in the transaction, with the roofer only left to apply an estimate of the extra slates required to accommodate changes in head lap.

The solution would be to sell stone slate in terms of the area which can be covered at 75 mm (3 in) constant head lap. If this is unacceptable, the industry should standardize sales in square metres of cover per tonne.

Promotion

Most companies recognise the need to promote their products, although for some (such is the shortfall in supply) this involves little more than maintaining contact with established customers. Local authorities' conservation departments contribute to the promotion of the product. As part of this policy, the Peak Park and Derbyshire County Council provide guidance on sources within and outside the region, and although at present these

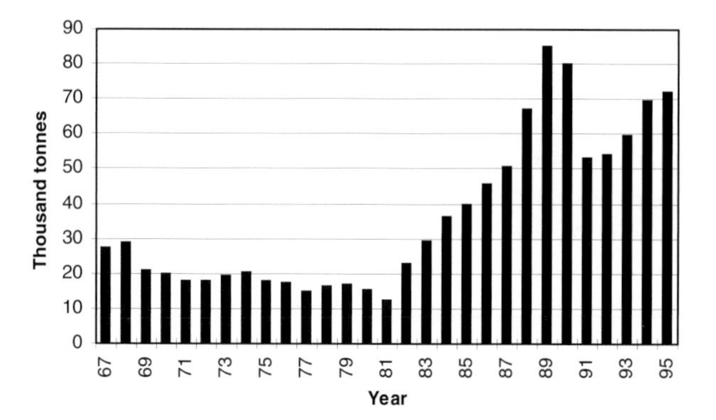

Figure 16. Sales of metamorphic roofing slates (after HM Customs and Excise and DoE Mineral Statistics 1997).

leaflets do not relate the source to the slate type this could be done if supplies of specific types become available (Derbyshire County Council 1996a; Hawkins 1992).

In a new initiative which has grown out of this study, a mobile exhibition has been developed explaining the problems stemming from the decline in the manufacture of stone roofing nationally, the work which is in hand to reverse this trend and promoting an increased understanding and use of the product. This has been displayed at the major trade and conservation events and is available to local conservation conferences and similar events.

Manufacturers should be encouraged to attend these events to promote their products. At least two companies have sets of marketing literature which show their product range and provide some technical information on strength and durability. One company promotes itself at stone exhibitions but none attend specialist roofing exhibitions such as Rooftech or the main construction industry event, Interbuild.

The cost of attending exhibitions is a barrier and could be especially so for a small manufacturer with a narrow product range or a single slate type and a small market. Cooperative promotion could provide a solution. The resurgence of the thatching industry over the last two decades owes much to the promotional efforts of COSIRA (the Council for Small Industries in Rural Areas, now part of the Countryside Agency) who developed a mobile exhibition and attended all the construction, roofing and conservation exhibitions. The Natural Slate Quarries Association (now the British Slate Association), which promotes the use of metamorphic roofing slates, has found it beneficial to share the cost of attending small exhibitions with an association display which includes the products of individual members. At the anticipated scale of the stone slate industry, this may be too much to ask but unless the industry's products are promoted, it is unlikely to grow to a size which can support the production of the full range of slate types.

A campaign, run either by the local authorities alone or with industry support, could be effective in publicizing the importance to the region of stone slates and ancillary quarry products. Based around press releases, this could be inexpensive. Care would be needed to coordinate publicity with supply so as to avoid stimulating a demand which could not be satisfied.

The use of the product already receives significant and effective promotion by both national and local conservation bodies through their survey and specification activities. Since these bodies fulfil the roles of promoters, users and 'quality controllers' for both the specification and construction of roofs, it would be most effective for manufacturers to focus promotional efforts directly at such organisations and at conservation events.

The survey and specification process is extremely important in supporting the market. However, there is a shortage of readily available information to help building owners, architects, conservation officers, builders and slaters to correctly specify and construct stone roofs. Derbyshire County Council and the Peak Park publish guidance on stone roofing and the reroofing process and this is currently being updated (eg Derbyshire County Council 1977, revised 1992, revised 1996). It includes a model specification which, when implemented, goes a long way towards ensuring that the appropriate materials are used and that the construction is correctly executed. Nonetheless, many architects and conservation officers feel that they are ill-equipped to fulfil their roles and often have to rely on advice from sources which are sometimes no better informed than themselves or not motivated by the best interests of conservation. This often results in disputes on site, especially over what is the correct or most suitable product, material, or technique and just what is actually possible.

If stone slates are to secure a market, it is very important to sort this situation out. Clearly, improved guidance and training are needed, as well as a system which ensures that competent and experienced slaters are always employed. The subject of training is dealt with separately below.

At present, there is no formal definition of product quality for stone slates, such as a British Standard, although there are both European and British Standards Institution committees which have such products within their ambit. As a consequence, the material's technical advantages cannot be promoted in a formal context but, with the low level of production and the apparent absence of unsatisfactory products in the market, this is not a handicap to sales. However, if supplies became available from other sources either locally or abroad, as has occurred with limestone slates, the need to be able to demonstrate satisfactory performance may arise.

If delves are brought into production, it will be important to ensure that their slates are durable. The lack of durability tests makes this difficult and it is therefore important that a study of the geological factors which influence durability should be implemented.

Product mix and profitability: the case for other products

It is the first principle of any quarry operation, and of sustainable development, that as much quarried rock as possible should be converted to products. Manufacturing slates alone may not be viable without a considerable subsidy which may be uneconomic or at an unacceptable level of profit.

The character of the region's landscape is that of its geology, not only in its dales and heaths but also the products of its rocks. Its stone walls and barns, houses, churches and bridges have all evolved into a distinctive style in harmony with the fields and moors. Stone slates, kerbs, flags and walling are often as distinctive and variable as the roofing. Already, it is possible to see repairs to field boundaries changing their character, as alien stone or differently shaped cresting is brought in from outside the natural source area. The replacement of stone flags or setts with concrete paviours is especially intrusive. Planning policies already recognise the need for authentic products, and the economic operation of a stone slate delph could provide the sources which would maintain the local distinctiveness.

There will therefore be an economic, conservation and market case for extraction to a greater extent than that solely required for slate, and for the production of ancillary items such as flagstones and other paving, kerbs and walling stone. None of these would need extra machinery beyond a simple hydraulic block-splitter. A case could equally be made for the production of sawn masonry, although this would require more sophisticated machinery and supplies of electricity and water.

Basis of competition

Within the stone slates market, competition hardly exists. With the exception of second-hand slates, it is entirely a seller's market, although neither building owners nor contractors will accept a price without resistance. Where stone slate must be used, the main, indeed the only, basis on which prices are held down is the availability of second-hand stone slates. This availability is not good and so has limited effect. However, even the vendors of stolen slates will presumably be under some pressure to dispose of their 'stocks' as quickly as possible and therefore may have to accept lower prices than they would otherwise. This places a baseline under the sale price of 'legal' stone slates.

Other products which may be substituted for stone slates include Welsh slates (sometimes new but more often second-hand), imitation stone slates (normally concrete) and plain concrete interlocking tiles. There is also substitution by clay tiles in the east and south of the region. Where the option to use such products is available, they exert strong price competition not only because they are cheaper but also because of the lower cost of the roofing labour involved.

Imitation tiles attempt to reproduce the appearance of stone slates, without success. As they weather, they support the growth of different vegetation to that which develops on sandstone or limestone and ultimately come to look very dissimilar to the real stone slate. Nonetheless, some consider them to be more acceptable than plain concrete tiles and planning authorities have sometimes accepted their use outside the most sensitive areas because of the unavailability of stone slates. However, any use of these ersatz products is detrimental to the re-establishment and sustainability of a stone slate industry.

The comparative costs for each of these products is given in Table 4. There is considerable variation in the price of reclaimed slates, depending on their availability.

Table 4. Comparative costs of roofing products (1995).

Product	£ per m^2 of roof
New stone slates	70
Reclaimed stone slates	33
Imitation stone slates	14-22
New Welsh slates	15-47
Reclaimed Welsh slates	15
Imitation Welsh slates	9-15
Concrete tiles	4-5

For manufacturers, there is some opportunity to avoid strict competition by alternating products: for example, roofing-thickness slabs are sometimes sold as interior flooring. This helps to support the price as activity in these two sectors of the stone market fluctuate.

Supply chain

Some sales are made to merchants, generally specialists in roofing materials (there are several established second-hand roof-material suppliers in the region), some to main contractors, but most to roofing contractors because they have the experience to be able to locate suppliers who have stocks.

TRAINING AND GUIDANCE

Manufacture

Manufacturers venturing into stone slate production for the first time are likely to have difficulty in finding experienced slate-makers. The craft may have entirely died out within the region, but it should be possible to find an experienced person from elsewhere to provide initial training. Once inducted into the basic skills, any quarry worker experienced in stone should be able to develop the necessary expertise, supported by visits from the trainer to maintain development.

Conservation and design

Architects, conservation professionals and other specifiers have expressed a need for training in the identification and selection of slates and in correct constructional methods. Roofing companies and merchants also have a role in the specification process and it is equally important that they should be well informed about the suitability of different slate types for particular situations.

For specifiers and conservation professionals, a one-day course would be sufficient to provide an understanding of the basics of slate types and construction methods. If this could be linked to both written guidance and sources of expertise to deal with the more difficult issues, most of the problems could be overcome. A prototype for such a one-day event was organised by the Society for the Protection of Ancient Buildings (SPAB) and Derbyshire County Council at the National Stone Centre at Wirksworth in 1994. This covered Southern Pennine roofing. It was repeated for Cotswold stone slating as a joint SPAB and English Heritage event in 1996. These events were to some extent an exploration of the problems currently faced by stone slating but the format is perfectly suitable for a more training-focused approach. With a suitable adjustment of the content and the development of written backup, this could provide all that would be needed for a basic grounding in the subject.

Where more detailed technical guidance is required, enquirers would need to be referred to individuals with the necessary wide experience in the field of roof construction as well as the use of stone slates. Such people are not common and since they are usually expected to provide advice without charge they are not always willing to give the time required to do so. This is a perennial problem in many industries and usually overcome by product manufacturers providing a technical service, either individually or through a trade association. However, it is unlikely that stone slate manufacturers would have the time or expertise to take on this role. The Stone Roofing Association was formed in 1998 to act as the point of contact between the manufacturers, specifiers and contractors (Stone Roofing Association, Ceunant, Caernarfon, Gwynedd LL55 4SA, UK. Tel: +44

Table 5. Proposed structure for NVQ Slating and Tiling.

Level 1	Basic non-craft skills
Level 2	Application of roofing materials to new work
A	Responsibilities of the specialist sub-contractor
	Fixed and variable gauge single lap tiles
	Fixed and variable gauge double lap tiles
	Artificial slates
	Associated weatherings. Stage 1
	Access and scaffolding. Stage 1
	Responsibilities of the roofing contractor for re-roofing
B	Re-roofing including stripping slates and plain tiles
	Associated weatherings. Stage 1
	Access and scaffolding. Stage 1
Level 3	Regular size natural slating
	Centre nailed random slating
	Head nailed random slating
	Complex and radial slating and tiling
	Shingles
	Artificial random slates
	Stone slating
	Associated weatherings. Stage 2
	Access and scaffolding. Stage 2

1286 650402; email: terry@slateroof.co.uk; www.stoneroof.org.uk). The need can best be answered by comprehensive written guidance, preferably in a form which relates stone roofing to the recommendations of BS 5534:1997 *Code of Practice for Slating and Tiling* and BS 8000:1990 *Workmanship on Building Sites Part 6 Slating and Tiling*.

At the time of writing, no university or college in the country included stone slating within its architectural courses. This clearly leaves the professionals on whom we rely for the conservation of historic buildings and the sensitive design of new buildings, ill-equipped to carry out this role. This omission needs to be addressed with some urgency. A suitable training module on the subject should be developed and made available to colleges by a visiting lecturer.

Roof construction

Training for slaters is even more important but much less easy to achieve. Currently, their training is provided by a number of colleges, although these do not all include random slating. A structure for National Vocational Qualification Assessment (NVQ) for slaters and tilers has been developed and is being implemented. This includes stone slating as one of the modules at Level 3 (see Table 5).

The inclusion of formal modules for random and stone slating is a very welcome development since there is a significant amount of bad stone-roofing practice evident, in the region and nationally. Its inclusion within the NVQ structure, while suitable for training apprentices, may fail to recruit the older roofer. A stand-alone course for experienced craftsmen is very desirable and could be developed from the NVQ structure and provided by roofing colleges.

Training providers

The Learning and Skills Council has responsibility for ensuring adequate provision of training. Their role in the field of stone slating would be focused on the training needs identified by the Construction Industry Training Board (CITB) and the funding of such provision by colleges.

Training in some rural crafts is funded by the Countryside Agency (formerly the Rural Development Commission), whose locus is to ensure that there is a rural dimension in national training policy. [6] It has developed a rural training model, the New Entrants Training Scheme (NETS), to meet the work-based needs of craft-based rural businesses. The NETS thatching apprenticeship, for example, covers water reed, long straw and combed wheat reed and includes a comprehensive book on the subject published by the Agency. This, together with the national standards set out in the thatching NVQs, should provide useful background for specifiers and conservation officers as well as roofers.

At present there are no new training courses proposed, but the Agency is working with other national training organisations, including English Heritage, to develop the NETS programme into a Rural Crafts Modern Apprenticeship.

THE PLANNING CONTEXT

In April 1993, a seminar was held in Matlock to discuss the resumption of stone-roofing production in Derbyshire and the Peak Park. This meeting assisted in formulating the brief for Phase 1 of the research. A second seminar was held in Bakewell in 1995. Among those attending were local authority representatives responsible for mineral planning, development control and the conservation of the built environment. English Heritage was also represented as the organisation responsible for advising the Department of Culture, Media and Sport on conserving the built environment and as a provider of grant aid towards the repair of historic stone slate roofs.

As a result of the seminars the following conclusions emerged.

Planning policies – mineral applications

The mineral planning authorities should consider whether their existing or proposed policies, or criteria for determining applications, require amendment to ensure that the planning process does not inadvertently constrain the opening of small quarries for dimension stone extraction.

The planning authorities should also consider to what extent it might be acceptable for a small quarrying company which specialises in stone slates to also carry out limited-scale manufacture of other items (such as concrete paving slabs faced in quarry waste) where this is a way to make stone slate production economically viable.

Planning policies – development control

Local planning authorities, especially the Peak Park, should review the practice of requiring, in some circumstances, the use of stone slate for roofing new buildings as this is stimulating the market in reclaimed stone slate, probably to the detriment of the county's existing traditional building stock. Consideration should be given to requiring the use of new roofing slate in such cases.

Historic Building Grant Aid policies
Grant-making bodies should consider whether or not grant aid should be adjusted to favour the use of new, as opposed to reclaimed, stone slates. Grant aid from local authorities, English Heritage and the Ministry of Agriculture, Fisheries and Food for the repair of stone slate roofs in Derbyshire inevitably fuels the reclamation trade and the cannibalization of repairable roofs.

Minerals planning

The Derbyshire County Council Minerals Plan (February 1996) is considered in detail here as an example. The policy within the Peak Park Minerals Plan includes essentially the same objectives and constraints, and:

- identifies production of building sandstone within the county as very small-scale, 5000–6000 tonnes per annum
- recognises that natural stone is sometimes needed as a building material for environmental reasons when projects are proposed in sensitive localities such as Conservation Areas and that this should be taken into account in the planning process
- recognises the importance to the built environment of the availability of local stone for the repair of historic buildings, field boundaries, paving, the construction of new buildings and, in particular, the production of stone slate roofing
- recognises that most sandstone reserves in the county lie within areas of high quality landscape, some being designated as Special Landscape Areas
- recognises that the need to safeguard the landscape will be a significant constraint on new proposals to work the reserves but recognises that, in practice, building-stone operations are often small in scale with modest production levels. This enables their impact to be minimised.

In view of the above factors, the policy for Minerals Planning 15 (Building Stone) states:

> Proposals for the extraction of rock for use as building stone will be permitted provided that:
> - the mineral authority is satisfied that there is a need for mineral of a specific character to be worked in that location
> - the proposal is small in scale and includes measures to ensure that its effect on interests of acknowledged environmental importance can be kept to an acceptable level.

Within the plan, there is also a preference for extensions to existing operations over new developments.

The mineral planning regulations permit the development on farms of quarries (without the requirement for formal planning applications) when the product is for use within and for that farm's own purposes. In view of the large number of quarries in the database which are situated on farms (and which presumably supplied the farm-buildings' roofs), the intermittent operation of such quarries could be very significant, especially within the great estates of the region. By using the mobile operation described above, they could well be made use of for even a single building repair.

The predisposition on the part of both of the region's mineral planning authorities to look favourably on applications to produce roofing slates is conditional on:

- the need for slates of a specific character to be demonstrated
- the scale of operations to be modest
- the impacts to be limited to an acceptable level
- there being a preference for production within existing quarries wherever possible.

Satisfying planning controls
The special issues that need to be assessed in determining an application for slate production, excluding those environmental and other impacts which would apply to most planning applications as a matter of course, are:

- the need for a slate of a specific character
- whether an existing quarry could supply the demand
- the extent to which production of other products might be permitted to ensure the economic viability of the enterprise.

The research identified seven generic slate types in the region (Table 2, Figs 9–15). Unfortunately, it has not been possible to quantify the market for stone slates in the area because of a lack of reliable and consistent information. In view of this, it is impossible to estimate the specific demand for any of the types but in total it will be small in comparison to other quarrying activities, including that of the very limited building-stone industry.

In 1999, only two delves were known to be producing slates within the region, at Kerridge (Kerridge type) (SJ 943757) and at Fulwood Booth (Yorkstone type) near Ringinglow (SK275853). The latter is very small scale.

However, it is acknowledged that demand for the Yorkstone type could be supplied from existing delves outside the region and that, if they can meet demand, this is preferable to new developments. Enquiries made during 1995 indicate that this supply might be variable. Manufacturers indicated low demand, citing competition from reclaimed material as a cause. Users complained of inadequate supply but normally would only use new material as the last choice, preferring cheaper, reclaimed slates.

The Magnesian Limestone of the Whitwell area (SK5576) has so few examples remaining that there is virtually no market for the product. Considering the very low demands for this slate, it is probable that there would be continuing demand for six types, one of which, Yorkstone, is available outside the region.

Since the supply of reclaimed slate is often from undesirable sources and one objective of this study is to eliminate this, it is anticipated that the demand for new slate of any type will grow. Therefore, even the Yorkstone type may require increased production.

There is a general paucity of supply of those slates which are most prevalent or most distinctive. Therefore, any applications for production of any of the types should be seriously considered if this will provide a supply which might satisfy an existing demand.

Extensions to existing delves
The mineral planning process would give preference to new or extended operations in an existing quarry with consent. All of the quarry operators listed in the database who were not producing slates were asked if they could do so. All said they could not or would not. Clearly, some of these had been slate delves in the past, as the remains of fissile rock could be seen at the surface of the workings, but present-day manufacturing objectives or methods make small-scale, hand-made production unattractive.

These companies either produce aggregates, where the extraction by fragmentation blasting is incompatible with slate manufacture, or masonry products. The latter delves have generally been developed into deep workings. Fissile rock almost always exists as a relatively thin surface layer which necessitates much larger extractive areas and hence more extensive planning approval. While this might seem unattractive to some companies, within existing operating plans it is not considered to be an insurmountable objection. Indeed one company, with access to fissile rock at a deep level, has recently commenced intermittent slate production in response to demand.

Where surface fissile rock does not fall within current production, it is envisaged that extraction could be subcontracted without interfering with the main operations. The worked areas could either be reinstated or left for future extension of the deep quarry. Since every quarry would eventually have to remove the overburden, this could provide considerable cost savings. The absence of slate production in an existing quarry with potential need not be an absolute barrier. This option should be given serious consideration by intending slate manufacturers.

The need to make other products
An economic, conservation and market case has already been made in this report for extraction to a greater extent than that required for slate, as well as for the production of ancillary items such as flagstones and other paving, kerbs, and walling stone. None of these would need extra machinery beyond a simple hydraulic block-splitter. Thus, approval of an application to exploit a deposit of stone slate should not necessarily be restricted to the production of that item alone. For reasons of economic viability and local needs for ancillary products, applications for production of these items should generally be regarded as a desirable concomitant of slate manufacture. The making of other items requiring more elaborate machinery and facilities should be judged on their own merits.

Even if all possible products are manufactured, there will inevitably be a certain amount of waste rock. This constitutes both an economic and environmental handicap to a manufacturer. Waste can be either:

- disposed of on site into old workings if they are available or onto the surrounding area
- carried away for disposal in a more acceptable way
- converted into a crushed product such as construction aggregates.

All of these options have an environmental impact, although tipping into old workings will probably be the least objectionable. Indeed, because of the bulking-up of broken stone, it can make a considerable contribution to the reinstatement of worked-out ground, albeit at a significant cost.

If unobtrusive disposal on site is not feasible, then crushing of waste may be an acceptable alternative. Using mobile crushers, it is likely to take place only at long intervals and for a short period of time. Inevitably, it will require more substantial access routes than the minimum required to operate the delph for stone slates etc. For small volumes of waste, it is unlikely to pay for itself.

Options for waste disposal have been assessed on both economic and environmental grounds.

Environmental impacts

Environmental policy
The policy of both Derbyshire County Council and the Peak Park Joint Planning Board is that proposals for mineral workings will be permitted, provided that their likely environmental impact is acceptable with regard to the effects (including visual and transport implications) on local communities, agriculture, water supplies, landscape, topography, wildlife, geology, the built environment, rights of way and recreation.

Applications to develop a delph must provide an environmental assessment (EA) to ensure that the likely effects of the development are fully understood and taken into account before it is allowed to go ahead. The full details are set out in EC Directive 85/37/EEC. The procedures and requirements of an environmental assessment are explained in *Environmental Assessment. A Guide to Procedures*, published by HMSO.

The regulations apply to two separate lists of projects: Schedule 1, for which an EA is required in every case, and Schedule 2 for which an EA is required only if the particular project is judged likely to give rise to significant environmental effects. The delving of stone slates falls within Schedule 2.

For Schedule 2, the significance of a development will essentially be assessed on the following criteria:

1 whether the project is of more than local significance, principally in terms of physical scale.
2 whether the project is intended for a particularly sensitive location, for example, a national park or a site of special scientific interest (SSSI), and for that reason may have significant effects on the area's environment even though the project is not on a major scale.
3 whether the project is thought likely to give rise to particularly complex or adverse effects, for example, in terms of the discharge of pollutants.

For a stone slate delph, criteria 1 and 3 are unlikely to be significant but criterion 2 may well apply, given the location of many old delves within the Peak District National Park or within areas of moor and heath.

The assessment criteria are amplified in Appendix 1 of the *Guide* and for Schedule 2 projects in the extractive industry specific reference is made to 'extracting minerals ... such as marble'. These will require an EA if they are likely to have significant effects on the environment by virtue of factors such as their nature, size, or location.

The developer's assessment of a project's likely effects is presented as an environmental statement. The content of this statement is set out in Appendix 2 of the *Guide*. It must contain, among other information, a description of the development's likely direct and indirect effects on the environment, explained by reference to its possible impact on:

- human beings
- flora
- fauna
- soil
- water
- air
- climate
- the landscape
- the interaction between any of the foregoing
- material assets
- the cultural heritage.

Detailed guidance on the assessment of the effects on each of these is given in Section 3 of the *Guide*.

Operational impacts

Because the demand for stone slates is never likely to be very large, their manufacture will inevitably be on a small scale and will have a correspondingly limited potential impact. To this extent, any proposed development should find it comparatively easy to satisfy the environmental constraints.

Manufacture is to a large extent by hand and is likely to remain so. Nonetheless, productivity will be as important a factor in the making of stone slates as for anything else. And so it should be, if producers are to avoid pricing themselves out of even a subsidised market. Consequently, mechanical handling should be adopted wherever feasible. This will be predominantly for topsoil removal and stacking, perhaps stripping of overburden, and waste removal. The need to remove topsoil and overburden may be infrequent: a delph operated for slate and flagging in Yorkshire carries out these processes once every few years. Alternatively, if a delph was operated for slate and flagging alone and extraction occurred near the surface, stripping and reinstatement could take place weekly or monthly.

The production process can be envisaged as:

- removal and stacking of topsoil (with or without overburden stripping) by back-hoe or face shovel, possibly by contract hire at infrequent intervals
- extraction of fissile rock by hand or winch
- hand splitting
- edge trimming and holing by hand or small machine
- packing for despatch by hand
- despatch by small truck or pick-up
- removal and tipping of waste by hand or small excavator
- reinstatement, possibly by sub-contractor or as a continuous process.

With the exception of the initial and final steps, it can be seen that this is more of a craft industry than what is normally understood as quarrying in the twenty-first century.

The general operation of a delph producing stone slates, walling and flagging will have an inherently low impact. The need for mechanical excavation and muck shifting will probably be intermittent and/or of short duration.

Transport requirements

The output of any delph of the type described in *Operational impacts* above will be quite modest. Based on the level of activity in similar delves which are currently operating, the day-to-day transport needs will probably involve no more than one or two journeys per day. They will, of course, depend on whether products are sold from the delph or removed to a central sales point. If sales are direct, there are unlikely to be more than ten visits from customers, even on a busy day.

On the same basis, it is estimated that transport for personnel and materials would account for up to ten return journeys per day. Transport of mobile plant to and from the delph would be very intermittent. Examples exist where it is as low as once every three years.

Agriculture

The delves envisaged will be quite small. Even if they were situated on high quality agricultural land, they would not take a large area out of production, perhaps only a few hectares. In fact, many potential delves are situated on land of poor agricultural value, heath, or moorland.

Water supplies

The effect on water supplies would need to be assessed individually. Potentially, water courses could be contaminated by fine particles of sandstone, diesel or other fuels and, if block sawing took place, by sawing additives. Otherwise, such delves are not users of chemicals. In practice, it is unlikely that delves will present a risk to water supplies or to the ecosystems they support.

Landscape and topography

Many disused delves occur in remote areas, often on the skyline. Consequently, they are prominent in the landscape. Unless a re-established delph is particularly well hidden by the lie of the land, it will certainly affect the local landscape, although on nothing remotely like the scale of the familiar aggregate and chemical limestone operations which are established in the region.

Given the range of slate types required and the other resource and economic constraints on a viable delph, it is unlikely that suitable potential delves will be found which do not have some effect on the landscape.

Visual impact

Historically, the sites in which slate delving has taken place appear to have been selected for convenience of extraction. This has resulted in the majority of delves investigated operating by one of three means:

- at the crest of a scarp slope with a face of about 10m height and worked along the strike. Cracken Edge (SK 037835) is a classic example of this form.
- in ground where the surface is parallel to the dip of the beds and close to horizontal. Extraction is by a single shallow bench or series of small shallow pits, with waste ideally being tipped into previously worked ground. Freebirch is one example.
- in steeper ground with the surface parallel to the dip of the beds. Extraction is by a series of benches working up-slope. The benches may develop into a series of distinct delves of modest height.

In the first of these cases, the delph will almost always be visible. In both of the others visibility will vary depending on the situation. Examples exist where the delph is virtually undetectable, either from close by or at a distance. In other instances, a delph cut into a hillside may only be seen from directly opposite.

In all newly opened delves, the disposal of waste may be obtrusive, especially in the early stages of operation, until and if it can be back-filled into worked ground. Reopening an old delph may provide an opportunity to tip into old workings.

Most of the workings visited in this study were remote from towns and villages, but delves do need access for vehicles. However, because the mobile plant and delivery vehicles need only be of modest size, such access routes would have no more impact than farm tracks.

Existing operational delves of the type under consideration do not use substantial buildings. One delph operates with only a shipping container which provides overnight storage space for equipment and a shelter during the day.

Wildlife

Where known, the wildlife interest of some old quarries is recorded on the database. Many are included because of their presence in general areas of significance, on the Moor and Heath maps for example, rather than for some factor specific to the quarry. The conservation policy of the 1994 National Park Structure Plan would confer a presumption against development other than in exceptional circumstances on any delph within a Moor and Heath area. Given the restricted occurrence of some slate types, their importance to the architectural heritage of the region and the small scale of a delph required to produce them, it may well be possible to argue that these are circumstances sufficiently exceptional to warrant approval.

Disused delves have the potential to be special habitats supporting rare or unique species, communities or ecosystems. Little work has been carried out within the region to assess the old delves individually, so specific guidance in this respect cannot be given here.

Geology

All delves add to the understanding of local and regional geology by exposing the rock, often over considerable distances. Special locations are protected by designating them as Regionally Important Geological Sites (RIGS) and these are noted in the database and the quarry records. It is possible that further old delves may be designated as RIGS as the work progresses. In any case, any planning application to redevelop a delph would be subject to such an assessment.

Recreation

Disused delves provide a location for many casual or organised recreational activities, whether approved or simply tolerated. Examples encountered in this study ranged from Country Parks managed by the local authority to casual picnic spaces. Activities included walking, mountain and trials biking, rock climbing and orienteering but generally these only occur in delves close to a road. By far the most prevalent activity is walking on footpaths, many of which are designated rights of way and probably originated as access to the delves they traverse or skirt. Some lie on the route of long-distance paths such as the Gritstone Trail.

The route through old workings may have been chosen because of its delving interest, so the re-establishment of a small-scale operation need not conflict with a footpath. Indeed, with appropriate interpretation, it may enhance the enjoyment of the trail.

Historical value of old workings

The working of sandstone generally, and stone slates specifically, has considerable historical importance in the region. The quarries may be significant features of the historical landscape.

Few examples of original buildings or machinery were found in this study. Those few buildings which were encountered were generally simple shelters constructed from waste rock.

Unfortunately, there is no context in which to assess the historical importance of specific sites as no research has been carried out in this field. It is possible that the database produced by this study may form the basis of such an investigation.

In the absence of any formal studies, it is the opinion of the field workers for this and the geological study that there is a prima facie case for conserving two delves, Cracken Edge (SK 037835) including White Rakes (SK 037843), and Glossop Low (SK 058964). The former is a RIGS, and together they represent two of the systems of extraction and were clearly of great commercial significance. Because they have stood virtually unchanged since their closure, they provide an opportunity to develop an understanding of slate-delving techniques. Teggs Nose (SJ 948725) is another site which has considerable historical and geological significance. It demonstrates two methods of extraction and has a large geological exposure showing a range of rock types of varying suitability for building products. It is already protected by its designation as a Countryside Park.

Sustainable development

Clearly, any delving has an environmental impact but the specific effects of stone slates are less than they might at first appear. Because they are largely hand-made and require little quarry plant, their production has very low fuel demand. Reliable figures are not available for the fuel consumption involved in making roofing material but they will certainly be much lower for stone slates than any manufactured roofing product, including concrete, resin or glass-reinforced cement imitations. The environmental impact is very small when spread over the life of the slates (at least 100 years and maybe several hundred).

From the above, it is clear that because delves manufacturing roofing, flagging, and walling products will almost inevitably be small-scale they should have a low operational or visual impact. The effects of transport on water supplies and agriculture will be slight or negligible. Impacts on recreational and social use, landscape, wildlife, and geological interests will be site-specific.

Conservation policy

Government views on the conservation of buildings are set out in Planning Policy Guidance (PPG) 15 *Planning and the Historic Environment*. This recognises the important contribution of groups of buildings, as well as individual examples of special merit, to local and national heritage and to the attractiveness of the communities in which people live, work and play. In the South Pennine region, stone slates and walling, flags and kerbs make a significant contribution to the appearance of individual buildings, as well as to extensive groups of buildings in towns and villages.

The Peak Park Joint Planning Board, Derbyshire County Council and the District Councils of the South Pennines contribute to the conservation of the region's architectural heritage by designating conservation areas, controlling alterations to listed buildings and, where possible, by assisting with the cost of repairs (in which English Heritage also has a role to play).

Although the principle of conservation is that roofs should be maintained and renewed in their original materials, in practice changes to roofs may only be controlled if they are listed or subject to an Article 4 Direction under the Town and Country Planning (General Permitted Development) Order 1995 (SI no.418). In conservation areas, consent for partial demolition may also be required.

Listed building controls relate to all works, both internal and external, which could affect a building's special interest. Consent is not usually required, except where alterations are involved which would affect the character of the listed building.

The substitution of one roof covering for another on a building which is not listed does not normally require planning permission. However under the provisions of either Article 4 (1) or (2), planning authorities can sometimes have the right of the property owner to do this removed, in which case such alterations become subject to planning control. The removal of permitted development rights is applied sparingly and generally only in conservation areas. At present, such controls apply only to about 10% of the conservation areas in Derbyshire and the ones affected are predominantly outside the stone slate areas.

Many stone slate roofs in the South Pennine region fall outside the control of listing and conservation areas. Most prominent among these are the field shelters and barns which contribute so much to the character of the countryside. These buildings are particularly vulnerable to dilapidation, because they often have no role in modern farming, to repair or replacement with inappropriate materials, and to theft, because they are often remote.

Redundancy of farm buildings is a significant factor in the changing appearance of the region. A national review in 1988 concluded that barns are the subject of more listed building demolition applications than any other building type (Darley 1988). Together with miscellaneous farm buildings, they account for 20% of all such applications.

The greatest prevalence of stone slate roofs is in the north of the study region, of which about half is in the Peak Park. Examples of such roofs are rare south of the Derbyshire Dales and Amber Valley District Council areas. The north's predominance in the total of Derbyshire's listed buildings is indicated in Figure 8.

Within the northern half of Derbyshire and in the Peak Park there are almost 100 conservation areas. In most of these, stone slates are a significant feature of the roofscape; in many, they predominate. There are also 1000 listed buildings with stone slate roofs in the study area and a further 2000 in the adjacent Macclesfield district, which historically drew stone slates from what is now the Peak Park.

The cost of carrying out repairs in stone slates is high, both for materials and labour. Recognising this, some local authorities provide grants for the roof restoration of individual buildings and English Heritage does so for listed Grade I and II* buildings and within Conservation Area Partnership Schemes (managed by local authorities). Roofs take the largest part of the building repair grant money available in the region. In some districts, it accounts for the whole amount available, while in others none is provided. It is generally agreed that the amounts available are inadequate and that consequently many stone slate roofs are not conserved. Under these circumstances, the alternatives adopted (generally concrete imitations) are entirely unsatisfactory. They are unconvincing imitations to begin with, then weather to a completely different appearance and their use is a deterrent to the revival of the stone slate industry.

The Countryside Stewardship Scheme (CSS) administered by the Ministry for Agriculture, Fisheries and Food has the objective, among others, of promoting the restoration of neglected land or features of the landscape including traditional farm buildings (those built before 1940), using methods and materials appropriate to their age, function and location.

This scheme also supports the conservation of field boundaries by management which follows traditional practices and uses local materials. There are two reasons why the old field boundaries of the region are built with stone. Firstly, they were a convenient place to put stones picked up from the fields. Secondly, stone was readily available as a 'waste' product from local delves exploited primarily for roofing, flagging and masonry. In this context, a local delph would frequently have been the one operated for the farm on which it stood or for the immediate community. The importance of walling stone as a by-product of roofing manufacture and the enhanced economic viability of such delves has already been emphasised.

The Countryside Stewardship Scheme which came into effect in April 1996 placed a priority on the conservation of dry stone walls and vernacular buildings within the moorlands and uplands of Derbyshire and the Peak Park. (The Environmentally Sensitive Area in the south-west of the Peak Park was excluded.) Under the scheme, grant aid (usually at 50%) can be provided. Listed buildings and buildings in conservation areas are included within the scheme, although work separately grant-aided by other sources such as English Heritage may not qualify for a CSS grant. Non-listed buildings have in the past received grant aid jointly from the CSS and the Peak Park Joint Planning Board or Derbyshire County Council.

Roof conservation in practice

Conserving stone slate roofs

The policy and practice for the conservation of roofs (as well as other aspects of buildings) is set out in the English Heritage publication *The Repair of Historic Buildings* by Christopher Brereton (1991, revised 1995) and within the guidance leaflet *English Heritage Conservation Partnership Schemes: Specification requirements for grant-aided work*.

Both documents highlight the need to salvage as many as possible of the existing slates and to reconstruct the roof to the same pattern of random sizes and diminishing courses as the original, as far as is feasible. It is recognised that there will inevitably be a shortfall of reusable slates and hence a need to obtain similar ones to make up the deficit. In this case, the preference is that 'these should be new (where quarries exist and appropriate slates can be obtained) or sound second-hand natural stone slates to match the existing in size colour and texture as closely as possible'. Brereton points to the need to avoid the use of cannibalized slates from other old buildings so that this practice is not carried out purely to supply the demand for slates. It is, of course, desirable that slates should be reclaimed if a building is being demolished or the roof removed for other reasons. The difficulty arises in supporting the latter while avoiding the former.

Nevertheless, in the absence of a reliable supply of new slates in the region, the normal practice is that second-hand slates are almost always used for both repairs and new work. This has led to a steady trade in these products, with many roofers being obliged to spend a considerable amount of time and effort seeking out suppliers. In these circumstances, it is hardly surprising that much of the supply comes from dubious sources, including wholesale thefts of roofs, some of which suffer this fate more than once. Isolated barns are particularly vulnerable.

Criteria for roof conservation

The appearance of a stone slate roof is an amalgam of many features: the texture, colour, size and format of the slates; the pitch of the slopes and the treatment of their intersections, and of the eaves, ridges, gables, verges, dormers and chimneys. Each of these features should be carefully recorded before restoration work commences, and modern techniques and materials should only be substituted where there is a sound technical reason to do so. The same traditional techniques and styles should be applied to new construction, extensions, and alterations. Guidance on the correct techniques for reroofing is provided by the conservation departments of Derbyshire County Council and the Peak Park Joint Planning Board. Currently, this guidance does not include criteria for the selection of replacement or additional stone slates. The following are suggested in order of importance to the roof's appearance.

- Rock type: sandstone should never be replaced by metamorphic slate or limestone. Aesthetic considerations apart, limestone placed above sandstone can be very damaging to the latter. Limestone dissolved in rainwater will percolate into the sandstone interstices, where it recrystallises as the roof dries, disrupting the sandstone and leading to spalling and early failure.
- Size range and mix: traditionally, these would have been a consequence of the characteristics of the rock and the policy in the delph. As far as possible, the same sizes should be obtained as were originally used. Because slates deteriorate at their top edges, reclamation and removal of delaminated top edges leads to a reduction in length. It is very important to obtain extra slates of the maximum length required in order to make up for this deficiency. Ideally, the pattern of reducing courses should be conserved, although there is a limit to what can be achieved without wholesale renewal of the slates. The length of exposed slate margins should be recorded before dismantling a roof. If the larger slate lengths can be obtained to make up losses due to delamination, then the diminishing coursing will usually work out to a close approximation of the original. It is important to realise that, in random slating, the head lap is not constant and must be increased at changes of length to ensure adequate resistance to driving rain. This influences the length of the slates' exposed margin but it must take priority over the need to conserve the roof's appearance.
- Thickness: this characteristic is very important because the edges give the authentic stone roof its bulky appearance. Its visual effect is influenced by the way in which the edges are dressed and the angle at which the edges spall during manufacture. Stone slates are sometimes turned over when being reinstated. The argument for this is that they have often settled into the roof, becoming curved. When replacing them, they have to be positioned with the concave face

downwards to avoid the tails kicking up. One disadvantage of turning reclaimed slates is that the edges will be bevelled the wrong way.

- Surface texture and grain size: surface texture may vary from completely flat and featureless to rippled and twisted. Besides the obvious effect grain size has on the appearance of the surface, it will also affect the build-up of surface deposits. Further, if the slate has an open structure, it will tend to absorb and hold more water, making it easier for lichens and other vegetation to become established.
- Colour and visible mineral: colour can be more significant than surface texture for the smoothest and flattest slates. When newly laid, the colour of slates is very evident but diminishes as dirt, soot or lichen growth build up. For this reason and because pigmenting minerals may be leached out, the colour of most old roofs is quite different to the rock from which they are made. Some minerals (such as mica) will give the surface of new slate a very distinctive appearance but this may alter with time. Little is known about how the colour and surface of stone slates change over the years and it may be that the soot-coloured appearance of many old roofs is becoming a thing of the past.

PHASE II – BUILDING THE FUTURE

Research aims

The Phase II research was undertaken with the assistance of a Partnership in Technology grant from the Department of the Environment and was completed in April 1997. This concentrated on the implementation of recommendations from Phase I, but was still largely centred around the South Pennines. Its main objectives were:

- to generate, and make publicly available, information to assist delph operators to identify environmentally suitable sites, to identify and meet market demands, product performance criteria and business plans, and to assist the re-establishment of production in the South Pennines and other regions
- to increase awareness of the importance of stone slate as a roofing material through promotional initiatives such as publicity campaigns, a travelling exhibition, and free leaflets
- to improve the level of awareness and understanding of conservation issues by producing technical guidance on stone slate roofing within Derbyshire, and a general guidance leaflet for building owners, specifiers etc, and encouraging other regions to follow suit
- to investigate available training for roofers and specifiers, and suggest improvements, including the introduction of training modules
- to promote the adoption of planning and other policies to facilitate the use of new stone slate as a roofing material
- to investigate and promote a marketing and sales system which discourages the use of inappropriate materials
- to give further consideration to the proposal for a central stock-holding production scheme in the region
- to promote the study as a model and encourage other stone slate areas to examine the difficulties in their own regions.

Product advice

During 1996, Terry Hughes, acting on behalf of the research team, provided a free support service to potential producers of stone slates both within and outside the South Pennines. This service is now available through the Stone Roofing Association.

The South Pennines

Since the study began, stone slates are now available from two new sources and a further one is progressing through planning. These are all in the north of the region and, to a large extent, only increase the availability of the Yorkstone type which is already manufactured in Yorkshire. However, one source may also have a match for the Freebirch type. Further south, about 20 enquiries have been received but progress is slow. An application to extend an existing quarry near Kerridge in the Peak Park to produce stone roofing was supported, but fell through when the applicant withdrew.

The Cotswolds

Support is being given to an initiative to restart production of Stonesfield slates in the region. These are frost-split and the project would have implications for Collywestons which are similarly made, as it seeks to develop means of artificial frosting for the manufacturing process to allow year-round production. A potential supplier has commissioned geological surveys in another area for a new slate quarry with the aim of increasing competition and offering edge-dressed slates. This quarry in the Forest Marble began production in May 1999.

Harnage, Shropshire

A proposal was made to fund and carry out desk and field research to identify a source of calcareous sandstone slates. These are urgently required for a listed church roof which needs replacement, and in the future will be necessary for a limited number of very fine historic buildings. Temporary work and reroofing was grant-aided by English Heritage. The intention is that this very small-scale operation could provide a model for the supply of similar materials in other areas where demand is insufficient to warrant a fully operational quarry. Delving was successfully carried out during the winter of 1998 and roofing was completed in 1999. (See 'Sourcing new stone slates and re-roofing the nave of Pitchford Church, Shropshire', this volume.)

Llanveynoe, South Wales

An assessment of suitability of the stone slates from two small delves in the Ochon valley in the Black Mountains has been carried out with English Heritage funding. They were agreed to be suitable for the building which prompted the investigation and are also a potential source for many buildings throughout Hereford, Worcestershire and Shropshire. The assessment was hampered by a lack of durability tests for stone slates.

Dorset
An assessment of supply and demand for Purbeck limestone slates is being carried out as part of the Phase II work.

Sussex
An assessment of supply and demand for Horsham sandstone slates is being carried out as part of the Phase II work.

Awareness-raising and training

Phase II of the research has largely focused on awareness-raising and training issues which were identified as key areas for development at the end of Phase I.

The Roofs of England: A celebration of England's stone slate roofing techniques
English Heritage launched *The Roofs of England* campaign in November 1997. The campaign includes a travelling exhibition, brochure and poster which place stone slate roofing in its geological, manufacturing and conservation context. The exhibition, available on loan to interested parties free of charge, has toured the local libraries in Derbyshire and appeared at a number of major exhibitions and trade fairs across the country, as well as at historic sites in England where stone slate roofing is significant.

Guidance leaflets
Derbyshire County Council has revised its advisory leaflets, which offer information on stone slate roofing in the South Pennines: *Derbyshire Stone Slate Roofs: Technical advice and model specification* and *Derbyshire Stone Slate Roofs: Guidance for owners of historic buildings*. Other local authorities in stone slate roofing areas (including Cotswold and Collyweston) are being encouraged to produce similar regional guidance based on the Derbyshire model.

To supplement these, English Heritage has published a Technical Advice Note outlining good stone slate roofing practice and examining traditional roofing methods.

Training roofers
A review of suitable craft training has been completed. There is no specific stone slate training available in colleges and English Heritage has been asked to provide guidance and support. The NVQs have recently been revised and include at Level 3 both traditional and stone slate roofing. However, there is concern that this is offered at too late a stage and many students do not complete their training to this level. This aspect is under discussion with the National Federation of Roofing Contractors (NFRC) and possibilities exist for aligning this with the Roofing Industry Alliance initiative headed by the Department of the Environment.

Training specifiers
A training module for specifiers/conservation professionals is in preparation and has been enthusiastically welcomed by a number of the universities which run conservation courses. A number of presentations outlining the project and recommendations have been made to conservation bodies and at events which focus on continuing professional development.

Conservation policy review
English Heritage is in the process of currently revising and strengthening its own policies to support the conclusions and recommendations of the research. These are outlined in the relevant section of this paper. Derbyshire County Council and the Peak Park Joint Planning Board are also examining their own policies with a view to aligning them with those of English Heritage.

The study as a model

The study has generated interest in other regions facing similar difficulties in relation to stone slates or other scarce traditional building materials. In late 1996, English Heritage convened a meeting in Stamford, Lincolnshire, which brought together local planning authorities and the Collyweston Stone Slaters Trust to discuss the problems experienced with the stone slate industry in the areas where Collyweston slates are used.

In Kent, the local authorities have also set up a preliminary meeting to examine the problems associated with the supplies of ragstone and are keen to use the South Pennine study as a potential model for improving supplies of this local material.

Over the next year, further work is planned to apply the model in other regions. At present, the priorities seem to be the Cotswolds and the Sussex area, where Horsham stone is needed for reroofing local buildings. The Clay Roof Tile Council has also been liaising with English Heritage to see how it could use this example to further its own aims.

CONCLUSIONS AND RECOMMENDATIONS

The industry

Manufacturing
There is almost no production of stone slates within the South Pennine region. The level of technology currently employed is satisfactory and, except for edge trimming (but not sawing) of the slates, there is little scope for mechanisation. Stone slates need to be hand-made because the source of their individuality is the craftsman's response to the character of the rock. The combination of hand and eye maximises the output of slates from unpromising material. Where it is possible to mechanise some stages of the extraction and conversion process it should be done without detriment to the product. The exceptions are sawing to thickness and size.

Market
On the basis of the information provided by roofing contractors in the study region and surrounding counties, there appears to be a market of at least 2000 tonnes a year in the locality and the nearby areas which it traditionally supplied.

Potentially, the market could be increased substantially if those roofs which have been replaced with concrete tiles and other alien products in the last 50 years could be returned to stone slates as they require renewal. This would recover the loss to the local heritage, reverse the trend of decreasing demand leading to declining supply and provide the basis to sustain a stone slate industry well into the future.

The revival of the industry may well need to be supported through its initial stages by special provisions. It is proposed that this can best be accomplished by eliminating the deliberate cannibalisation of old roofs and by directing roofing grant money more directly to the benefit of the producers. Manufacturers may need encouragement to build stocks for a small market. This objective could be achieved by the creation of a central stock-holder.

Marketing

Position

Stone roofing has a good technical and aesthetic reputation. Manufacturers should position the product in the top end of the market, emphasizing its beauty, durability and historical and conservation importance. Its durability is an especially strong counter to the perceived penalty of the high initial cost.

Price

The price of stone slates is undoubtedly high and is unlikely to fall significantly. However, life-cycle costs are very much lower and this aspect should be promoted. The system of pricing stone roofing is complex and confusing. A simpler standardised method is recommended, based on the cost per square metre of roof at a standard head lap of 75 mm (3 in).

Profitability

Delves could be economic if they produced flagging, kerbs and walling, for architectural, conservation and environmental reasons, as well as roofing slates.

Competition

Almost every slate or tile exerts some competitive pressure on stone slates, generally by cost. The main competition for manufacturers is from second-hand stone slates and concrete imitations. Conservation strategies should emphasise lifetime costs and long-term appearance, and reinforce the need for like-for-like replacement in stone slates, or reinstatement of them when imitation products fail.

Promotion

Currently, product promotion tends to be reactive. If the market for stone roofing is to develop in the future, it will need to focus on the advantages of the material and publicise its range through trade shows and literature in order to counter the significant initial cost advantages of all the alternatives.

The use of stone slates receives considerable promotional support from a variety of conservation organisations, both regional and local. This support has included workshop events explaining the products and their use; these have acted as a point of contact between manufacturers and users. It is recommended that such activities should be developed jointly between the conservation bodies, the manufacturers, and the roofing industry.

Training

There is a pressing need for training in the manufacture of stone slates, their identification and the specification and execution of stone slate roofs. The lack of slaters skilled in the use of stone slates is a serious threat and it will certainly get worse unless action is taken soon.

The planning environment

Minerals planning

There is a predisposition on the part of both the mineral planning authorities in the region to look favourably on applications to produce roofing slates. In view of the general paucity of supply of those slates which are most prevalent or most distinctive, applications for production of any of the types should be assumed to supply an existing demand.

The absence of slate production in an existing active quarry with the potential to do so may be due to a disinclination on the part of the present operators to change their working methods. Stone slates are viewed as a difficult product. The option to manufacture stone slates in an existing quarry should be explored by prospective manufacturers before considering opening or reopening a new or disused one.

An enterprise based solely on stone slate production is unlikely to be economically viable. Approval of applications to exploit a deposit of stone slate should encourage the manufacture of ancillary products such as walling, kerbs, flagging and other paving. Manufacture of other items requiring more elaborate machinery and facilities should be judged on their own merit.

The options for waste disposal should be assessed on both economic and environmental grounds.

Environmental impacts

The general operation of a delph producing stone slates, walling, and flagging will have an inherently low impact. When amortised over the full life of the products, which may be 200 years, the environmental cost is low. Delves' need for mechanical excavation and muck shifting will probably be intermittent and/or of short duration, and they are likely to have a low operational or visual impact, except where they are located on the skyline. The effects of transport on water supplies and agriculture will be slight or negligible. Impacts on recreational and social use, landscape, wildlife, and geological interests will be site-specific. Energy consumption is probably lower for the production of stone slates than for any other roofing material.

Little work has been done to assess the specific wildlife value of old delves, although many are in the Moor and Heath areas in which there is a presumption against development. The geological importance of some delves has been assessed and a small number have been designated as Regionally Important Geological Sites.

There has been no comprehensive study of the history of the stone slate industry. This is surprising and leaves a significant gap in our understanding of the social and industrial development of the region and its vernacular architecture. It is desirable that such a study should be undertaken, not only because of the historical and social interest but in order to provide a context in which to assess applications to re-open old delves.

Architectural conservation

It is important to recognise and conserve the regional characteristics of stone slate roofing. In many areas, a wide variety of rock types has traditionally been used and new stone slates should match the originals as closely as possible in terms of geological type, colour, texture and thickness. Existing features of old roofs should also be retained, such as the range of slate sizes, and verge, ridge, valley, eaves and abutment details. Within the study region, there is a variety of slates; seven generic types would reasonably cover the historic range in the South Pennines (Table 2).

The types are not restricted to their type locality. Some may occur widely but intermittently. Their use outside their own type area, although largely unavoidable at present, results in a loss of local distinctiveness. Because many local slate types are not manufactured at present, policy emphasis on the use of authentic materials unintentionally promotes the cannibalisation of other stone roofs to the detriment of the regional roofscape. Wherever possible, stone slates supplied for the repair or restoration of historic buildings should be new in order to prevent this. Second-hand slates should only be used on the buildings from which they were recovered.

The delves

A database of 167 delves has been created. As far as was possible, each entry records the presence or absence of rock suitable for stone slate production and other geological, access, and impact information. In most cases, there is little information available on the historical, geological, and biological importance of the delves, so the significance of these factors would have to be included in an environmental impact assessment for each site.

Photographic records have also been made of delves, slates and roofs. Together with a library of stone slates which is currently held at the National Stone Centre at Wirksworth, they provide a record of the historical range of stone roofs in the region.

A review of the geology of the county with reference to the localities of fissile rock is provided. The database and the review can be used to short-list potential delves for re-opening or to locate likely sources of suitable fissile rock on new ground.

Further research

Further research should be undertaken into the history of the stone slate industry, roofing techniques and practice and the associated technical issues. It is desirable that such studies should be undertaken not only because of their intrinsic historical and social interest but to provide a context in which to assess applications to re-open old delves.

There are presently no means of comparing stone slates in technical terms against one another, as there are for metamorphic slates, and they are not covered by a British Standard. Production attempts stimulated by this research work have frequently tried to carry out testing to prove the durability and quality of the products, largely without success. Research is needed to establish suitable test methods so that comparative and base data can be provided for stone slates.

This study started at a time when the industry was in decline but there are many reasons to be optimistic about its future. The slate rock is available in abundance. There is a willingness to find solutions to the problems which the industry faces. The skills to make the slates and to create the roofs do still exist. Perhaps the most important asset is that there are people who care about stone slates and their importance to the character and beauty of the region. They can make the difference. What are needed now are enthusiasts to restart the industry.

ENDNOTES

1 A summary of the report prepared by Pat Strange of the Derbyshire Historic Buildings Trust, which appears in full in Volume I of Hughes 1996, is included on page 3, 'The Historical Context'.

2 A full copy of the report (Hughes 1996) is available from the Building Conservation & Research Team of English Heritage. The database is available for consultation at the offices of English Heritage and at any of the Derbyshire County Council Libraries.

3 The comment that the stone eaves-slates were used to protect tiled roofs is probably incorrect. They were more probably the remnant of a stone roof progressively replaced with tiles during successive reroofing or used to span to the outer face of the wall from rafters set onto the inner face.

4 Since the study was completed another roof has been discovered and the stone slate delph identified at Bakestone Moor, Whitwell.

5 A listing of stone slates for buildings in the database in Hughes 1996, vol 1, produced the distribution within the county. It does not include any roofs which have been entered as flags and so may be a slight underestimate.

6 Since this report was prepared the Countryside Agency intend to sub-contract most of the training to specialist colleges.

BIBLIOGRAPHY

Alcock N W, 1981 *Cruck Construction: An Introduction and Catalogue*, CBA Research Report 42, London, Council for British Archaeology.

anon, 1659, (1945) *The Art of Building or an Introduction to All Young Surveyors in Common Structures*, facsimile edition, Farnborough, Gregg International.

Beresford G, 1975 *The Medieval Clay-land Village: Excavations at Goltho and Barton Blount*, Society for Post Medieval Archaeology Monograph **6**, London, Society for Post-Medieval Archaeology.

Brereton C, 1991 revised 1995 *The Repair of Historic Buildings. Advice on principles and methods*, London, English Heritage.

Brunskill R, 1978a Distribution of building materials, *Transactions of the Ancient Monuments Society*, NS **23**, 41-65.

Brunskill R, 1978b *Illustrated Handbook of Vernacular Architecture*, London, Faber.

Craven M and Stanley M, 1991 *The Derbyshire Country House*, Derby, Beedon Books.

Darley G, 1988 *A Future for Farm Buildings*, London, SAVE Britain's Heritage.

Derbyshire County Council, 1977, revised 1992, 1996 *Derbyshire Stone Slate Roofs, guidance for owners of historic buildings*, Derby, Derbyshire County Council.

Derbyshire County Council, 1996a *Derbyshire Stone Slate Roofs, Technical advice and model specification*, Derby, Derbyshire County Council.

Drage C, 1990 Dale Abbey: the South Range excavation and survey 1985–87, *Derbyshire Archaeological Journal*, **110**, 75.

Durant D and Riden P (eds), 1980 *The Building of Hardwick Hall: 1* Derbyshire Record Series **4**, Chesterfield, Derbyshire Record Society.

Durant D and Riden P (eds), 1984 *The Building of Hardwick Hall 2: The New Hall, 1591–98*, Derbyshire Record Series **9**, Chesterfield, Derbyshire Record Society.

English Heritage, 1998 *Stone Slate Roofing*, Technical Advice Note, London, English Heritage.

Farey, 1811, reprinted 1989 *A General View of the Agriculture and Minerals of Derbyshire*, vol **2**, Chesterfield, Peak District Mines Historical Society.

Glover C and Riden P (eds), 1981 *William Woolley's History of Derbyshire*, Chesterfield, Derbyshire Record Society.

Hawkins R, 1992 *Sources of Building Stone for Use in Derbyshire*, Derby, Derbyshire County Council.

Hughes T G, Thomas I A, Guion P D, O'Beirne A M and Watt G R, 1995 *Roofing Stones in the South Pennines*, unpublished report for the Department of the Environment London.

Hughes T G, 1996 *The Grey Slates of the South Pennines: A report into the potential to re-establish the roofing slate industry of the region. Volume 1 The Industry, Volume 2 The Quarries and the Slates*, London, English Heritage, Derbyshire County Council and The Peak District National Park Planning Authority.

Innocent C F, 1916 *The Development of English Building Construction*, Cambridge, Cambridge University Press.

Kerry C, 1901 The Court Rolls of Baslow, in *Derbyshire Archaeological Journal*, **23**, 4.

Mercer E, 1975 *English Vernacular Houses*, London, HMSO.

Morris W, 1890 *On the External Coverings of Roofs*, London, SPAB.

Munsell Soil Charts, 1992 Newburgh, NY, Kilmorgan Instruments Corp.

Platt C, 1994 *The Great Re-building of Tudor and Stuart England: Revolutions in Architectural Taste*, London, UCL Press.

Stanley M G, 1993 *The Stone Slates of Derbyshire C: An Appraisal of Farey's List of Quarries*, unpublished, Derbyshire County Council.

FURTHER INFORMATION

To borrow the exhibition *The Roofs of England*, obtain brochures or posters, or for assistance or advice regarding potential production, planning or student dissertation topics, please contact: Terry Hughes, Slate and Stone Consultants, Ceunant, Caernarfon, Gwynedd LL55 4SA, UK. Tel: +44 1286 650402; email: terry@slateroof.co.uk

To obtain Derbyshire County Council's leaflets contact: Environmental Services Department, Derbyshire County Council, County Hall, Matlock, DE4 3AG, UK. Tel: +44 1629 580 000.

Stone Roofing Association website: www.stoneroof.org.uk

Please note that English Heritage is unable to provide a technical advisory service to the general public. For further local information regarding local technical advice, roofers and supplies of stone slates, please contact the Conservation Officer at your local planning authority or the Stone Roofing Association, Ceunant, Caernarfon, Gwynedd LL55 4SA, UK. Tel: +44 1286 650402; email: terry@slateroof.co.uk; www.stoneroof.org.uk

AUTHOR BIOGRAPHIES

Terry Hughes worked in the Welsh slate industry at Penrhyn Quarry for 10 years. He also has experience of the slate industry in the USA, throughout Europe and in India. Since 1983 he has represented the British slate industry on the British Standards Institution's committees for roofing slates (as chairman), the code of practice for slating and tiling and the National Federation of Roofing Contractors' slating and tiling committee. He has also been the chairman of the European (CEN) committee for slate and stone roofing products for 14 years. Since 1993 he has been an independent consultant. As English Heritage's consultant on slate and stone roofing he has carried out studies into stone roofing throughout England and has advised on the conservation and construction of the roofs of many historic buildings. As part of his work to support the regeneration of stone slate delving he established and chairs the Stone Roofing Association and the Stone Roof Working Group.

Susan Macdonald graduated as an architect from the University of Sydney and completed her conservation training at ICCROM, Rome. Susan has worked in private practice in Sydney and the UK. After five years with English Heritage, she is now Principal Heritage Officer, Local Government Heritage Management at the New South Wales Heritage Office in Sydney.

Patrick Strange is a founder Trustee of the Derbyshire Historic Buildings Trust and a member of numerous other bodies concerned with that county's historic and built environment. Now retired from teaching electrical engineering in the University of Nottingham he still lectures on the applications of geophysics to archaeology and has carried out geophysical surveys at many sites throughout the British Isles and Europe. He runs his own consultancy on historic industrial buildings and is currently archaeologist to the Cromford Mill Project, site of Richard Arkwright's first water-powered cotton spinning mill of 1771, which is currently undergoing a £2m restoration and development scheme.

Chris Wood is a senior architectural conservator in the Building Conservation & Research Team at English Heritage. He is responsible for managing several research projects and providing technical advice on remedial treatments to deteriorating historic fabric. He managed the English Heritage Master Classes at Fort Brockhurst, (now at West Dean College) which specialised in the 'hands-on' repair of ancient monuments. Previous experience was gained as a director of a private sector architectural practice specializing in building conservation, and as a conservation officer in local authority planning departments.

Stone roofing in England

TERRY HUGHES
Slate and Stone Consultants, Ceunant, Caernarfon, Gwynedd LL55 4SA, UK.
Tel: + 44 1286 650402; email: terry@slateroof.co.uk

Abstract

Following the study of the stone slate industry and market in the South Pennines, the situation for the rest of England was researched. The historical sources of stone slates and their geology are reviewed and the challenges to the continued regeneration of stone slate delving are discussed. The current state of supply and demand in the conservation of stone slated buildings is assessed.

Key words

Geology, quarry, quarrying, stone slate, delph, delving, tilestone, flagstone, sandstone, limestone, roofing, building conservation, vernacular architecture, building crafts

FOREWORD

This study, originally written as a report for English Heritage, has attempted to do two things. Firstly, to assess the supply and demand for stone roofing [1] in each of the main stone slate regions of England, and, secondly, to bring together in one document much of the information about the various stone slates and their historical sources which is dispersed in a number of published and unpublished documents and is often difficult to find. The detailed research was carried out between 1996 and 1998 but was updated in 2001 as far as possible to take account of subsequent events or new information.

The number of stone slates is almost limitless, depending on the visual, geographical and geological criteria used to define them. To try to keep the information manageable and accessible this study has been structured around a combination of geographical and geological criteria which are, in effect, markets for specific stone slate types or a group of similar types. Inevitably, such a structure cuts across county and regional boundaries. Gloucestershire, for example, appears in two sections: the Cotswolds group, which covers the Jurassic limestone sources, and the Welsh Marches which includes all the other, sandstone, sources. While this is inconvenient for anyone working in the Gloucestershire region, this approach avoids repeating the Jurassic information for ten counties from Dorset to Yorkshire.

Assembling the information on which this report is based has not been easy. Information about markets is largely unavailable: often no-one knows how many stone-roofed buildings there are in a region, especially farm outbuildings, or, if the records do exist, they are not easily accessed. Even where basic information on the number of stone roofs is available it was not possible, within the resources of this study, to make an assessment of how many will need re-roofing in the short to medium term. In addition, no data exist on the number of buildings which have been recovered in other materials, but which are now again in need of re-roofing and could return to stone slates if supplies became available.

Accurate statements about the geological sources of stone roofing have also caused considerable difficulty. Modern geological memoirs largely ignore these products and to research their location both geographically and stratigraphically, it was usually necessary to go back to records from at least the nineteenth century. However, more modern detailed geological mapping has resulted in modifications to the designation of some of the strata from which stone slates were obtained. As a result the names of some geological units have changed since the original references were written. To avoid others having to repeat the exercise of finding sites from early references and relating them to their modern stratigraphic name, many of the references have been reproduced in full. Even so, care should be exercised when using them to ensure that their names and stratigraphical designations are current.

In this report, the term 'stone slates' is used, rather than 'tilestones', which is preferred by some geologists. However the word tilestone does appear in some quotations from geological texts and Tilestones is used when speaking specifically about this named formation in South Wales and the Marches. Elsewhere, particularly in the north, stone slates are known as grey slates which distinguishes them from the 'blue' slates which, in this region, means metamorphic slates of any colour. 'Grey slate' is used occasionally in this study. A further term may be unfamiliar: delph, an old word for quarry (see Introduction). Another tradition continues in the stone slate industry and in stone roofing: the use of feet and inches. This convention has been followed here, but has been converted into metric units where it is sensible to do so.

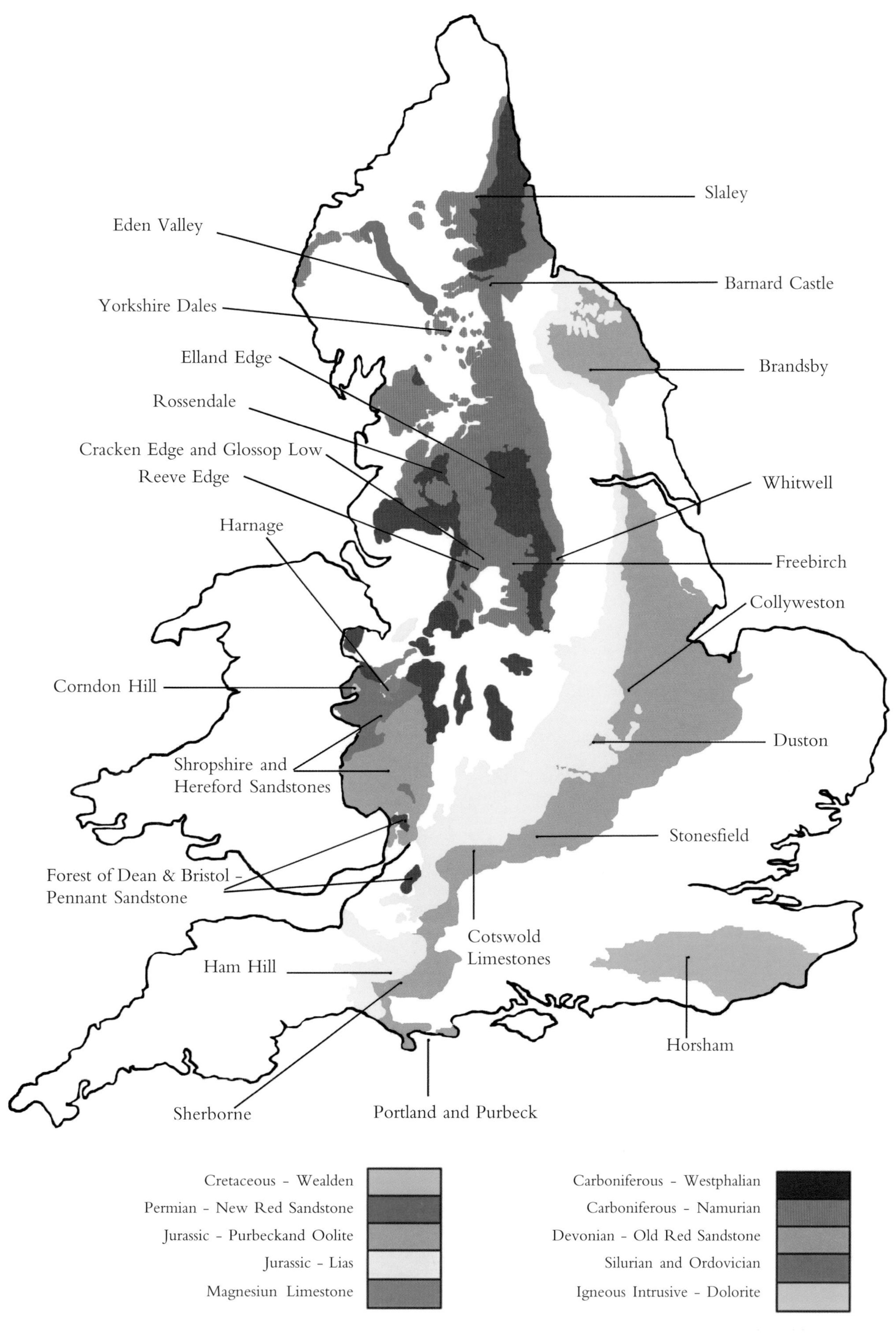

Figure 1. English stone slate geology and some historical sources (© Terry Hughes, after the British Geological Survey).

INTRODUCTION

> Saxum laminosum 'Tis call'd commonly Slate merely because 'tis us'd and indeed very fitly, like Slate, for the covering of Houses, particularly at Bath and in several Parts of the West. But it will not split, as Slate does, being found form'd into what they call Flaggs, or thin Plates; which indeed are no other than so many Strata. I have observ'd of them, betwixt Castleton and Worksworth, in the Peake of Derbyshire, and in some other Places, from the thickness of paper, thro' all Degrees to a very considerable Bulk. They increase desending, the thickest lying very deepest in the Earth. (Woodward 1728, 7–10)

There are few durable stones which have not been exploited for roofing, provided they can be split to a reasonable thickness, generally not more than one inch thick (25.4 mm) although there are exceptions even to this heavy limit. With the exception of West Sussex where Horsham Slate is used, the main stone slate regions sweep up through England from Somerset and Dorset to Cumbria and Northumberland. They include all the counties of the Welsh Marches and extend into Wales, especially in the Silurian, Přídolí series (at the boundary of the Silurian and the Old Red Sandstone), the Devonian Old Red Sandstone and the Carboniferous Pennant formation. The Jurassic limestones are important from Dorset to Northamptonshire. In the northern counties Carboniferous sandstones predominate, and near Lazonby in the Eden Valley of Cumbria and north of Dumfries, the red sandstones of Permian age have produced distinctive roofs (Fig 1).

> Fissile rocks. Bandy metal; Beechleaf marl; Boards, boardstuff, boardy clift; Book clay; Carbie; Clay-slate; Cliff, clift; Clives; Corrity; Delvin; Eddy rock; Fakes, faikes, fekes; Flag, flagstones; Flakes, flaikes; Flue; Killas; Leaf clay; Leafy post; Leats; Linsey, linstey; Metal, metalstone; Metal till; Pavey; Pendle; Peters; Plank, planking; Plate; Rattler; Riliff; Sclit, sclutt; Shab; Shales; Sheerbate stone; Shellat, shillet; Shill, shillet; Shindle-stone; Shiver; Skerry, skerrystone; Slab-stones; Slate, slatt, slatestone, slate ground, slatter, slattern, slaty band; Slig, sliggeen; Sliving; Tilestone; Tilly freestone; Wood coal; Wool. (Arkell and Tomkieff 1953, 153)

Roofs can be made with any fissile stone, but the term stone slate is only used for those which are sedimentary: mainly limestones, sandstones or, in a few instances, igneous. They do not include real slates, which are metamorphic. It is important to realise that although stone slates are usually described as limestones or sandstones the distinction between them is not precise and many 'sandstones' are calcareous and 'limestone' slates are often produced from localised sandy beds.

The names used for stone slates are usually either the geological formation in which they occur or the place where they are found. However, the formation name does not necessarily describe the stone or its properties. For example, the now obsolete term 'New Red Sandstone', introduced in the nineteenth century, includes both Permian and Triassic strata, which as well as red sandstones also contains magnesian limestones, marls and evaporite.

History

The use of stone for roofing has a very long history. Archaeological studies have recorded their use at Roman sites throughout England and Wales and there are examples in many rocks (Table 1). These include Purbeck limestone at Encombe and Norden near Corfe Castle, Collyweston Slate at Irchester and Apethorpe[2] and Cotswold stone slates at Ditchley and Shakenoak. In the sandstones Pennant has been frequently recorded in the Bristol region, and was used at Roman sites at Llantwit Major and Ely near Cardiff. Further north a micaceous sandstone roof tile was excavated at Uriconium B, now Wroxeter, south-east of Shrewsbury.

It appears that stone slates fell out of use with the departure of the Romans and the earliest subsequent records come from the thirteenth and fourteenth centuries. By this time they were being used in many parts of

Table 1. Geological sources of stone slates in England.

Region of use	Sources
Dorset and Somerset	Purbeck, Forest Marble, Upper & Lower Lias
Wiltshire, Gloucestershire, Oxfordshire	Purbeck, Corallian, Forest Marble, Hampen Marly Beds, Taynton and Trougham tilestones including Stonesfield Slate, Chipping Norton Limestone, Lower Fullers Earth, Inferior Oolite, Marlstone
Northamptonshire, Rutland and Lincolnshire	Blisworth Limestone, Rutland Formation, Collyweston Slate, Duston stone
Surrey, Sussex, Kent	Horsham Stone, Chatwell Stone
Gloucestershire, Bristol region, South Wales	Pennant Sandstone, Tilestones, Old Red Sandstone, Lower Lias?
Welsh Marches	Old Red Sandstone, Tilestones, Cheney Longville Flags, *alternata* Limestone, Chatwall Stone, Harnage Slates, Grinshill Flagstones, Corndon Hill dolorite
Staffordshire, Cheshire, Derbyshire, Lancashire, Yorkshire, County Durham, Northumberland	Carboniferous Millstone Grit and Coal Measures sandstones
East Derbyshire, Nottinghamshire	Magnesian Limestone
North Yorkshire	Carboniferous Millstone Grit and Coal Measures sandstones, Jurassic limestones
Cumbria	Carboniferous Millstone Grit and Coal Measures sandstones, Permian (New Red) Sandstone

the country. Accounts of their use are recorded by Thorold Rogers (1882), Walton (1941), Salzman (1952), Aston (1974), Lawson (1985), Moorhouse (1990), Hughes (1996) and Baldwin (1998).

In Purbeck the quarries which had been worked for masonry from an early date also produced roofing. In 1447 a parcel of white Purbeck slate was bought for 20 shillings (Thorold Rogers 1882).

In 1238, Cotswold stone was being exploited for roofing at Woodstock, although, contrary to many opinions, this would have been the surface weathered 'presents' rather than deliberately frost-split Stonesfield slates (Aston 1974). In 1250, Kirkstall Abbey used the Elland Flags of Yorkshire (Walton 1975, 40 quoting Mayhall [1861], 127) and, by 1286, 'sclatestone of Peterborough', possibly an early reference to Collyweston Slate, was used to roof Cambridge Castle. [3] The use of Collyweston Slate was well established by this period with records of 14,000 slates being supplied to Rockingham Castle in 1375 and 5,000 to Oakham Castle in 1383 (Arkell *et al* 1947). Thorold Rogers recorded the use of what must have been mainly stone slates throughout the Midlands from 1410 to 1570. He described how the size varied: 'the three kinds generally discovered in the Oxford accounts being that of common large, middling and small'. He also noted that they were 'sometimes bought after they had been shaped and holed, called *bateratio*, sometime in the raw state' (Thorold Rogers 1882, 435).

By 1367, the Hoar Edge Grit was being worked near Acton Burnell, south of Shrewsbury, and had been used at Harley to the east. In 1489 even such a remote site as Corndon Hill, in Montgomery (now Powys), was exploiting dolerite for roofing. This would later be used at Pride Hill in Shrewsbury (Lawson 1985, 116). Further south, at Stanton Lacy near Ludlow, roofing supplied in 1390 was probably from the Old Red Sandstone. The Cretaceous sandstone of Sussex was also exploited from an early date. In 1301, 2,500 'stones called scletes' were transported from the Shortsfield quarry to Thorney near Horsham. The Wiston Rolls also record the carriage in 1357 of 12 wainloads of stone from Horsham to Wiston and the payment of 3d to 4d per day for the roofers (Sussex Archaeological Collection 54.152). A hundred years earlier stone for roofing was being worked at Abbey Dore in Herefordshire.

> Those present and future should know that I, Hugh son of William le Crone of Moccas, have given and conceded for myself and my heirs, and confirmed by this my present charter, to God and St Mary and the monks of Dore, serving God and forever to serve him there, in free, pure and perpetual alms, all that piece of land with all its appurtenances and liberties, which lies between the land of these monks and the land of Margery le Crone, my mother, as appears by the marks and boundaries placed between them, and which are named before in the charter which I have from the same Margery. To have and hold to themselves and their successors from me and my heirs fully, freely, peacefully, and quit of all secular service, suit of court and claim which belong to this land, or could belong in any way. Also I concede and give to these monks the marl, sand and also slate, and the quarry which I have by the gift of the same Margery, my mother, in all her land wherever they can be found, with free entry and exit, without injury or any hindrance or forbidding, and also common of pasture throughout all this land. And if anyone wishes to injure or hinder these monks over these matters, I and my heirs will acquit and maintain them through all things at our own expense against the chief lord [of the fee] and whatever others, in all things and for ever. And so that this my concession and gift should endure valid and undisturbed forever, I have strengthened the present writing by the application of my seal.
>
> With these witnesses: Lord Walter, vicar of Bredwardine, Roger son of Hugh of Radnor, Richard le Breth, Peter the clerk, John de la Bache, Hugh de la Bache, John Muschet and many others.
>
> Given at Dore on the day of Saint Ethelbert, king and martyr, 56th year of the reign of Henry III [20th May 1272]. (Hugh, 1272)

In the north of England, Stephen Moorhouse's study of the Court Rolls and other documents relating to 20 locations in the West Yorkshire Coal Measures revealed an extensive industry in existence between 1314 and 1524 (Moorhouse 1990). Salzman (1952, 393) also recorded that John Fossor, Prior of Durham 1341–74, discovered a 'quarry of sclatstane at Beaurepaire' [4] and the execution of a building contract, including stone tiles for the roof, at Brandsby in 1341. If this is Brandsby in North Yorkshire it may be an early reference to the exploitation of the 'Great oolite, about Brandsby in the Howardian Hills [where] the grey limestone series yields a hard siliceous limestone, from which ... large slabs and roofing tiles were obtained' (Howe 1910, 323).

Contrary to the generally held view that the early use of stone slates was restricted to important buildings, his review of documentary evidence in West Yorkshire led Moorhouse to conclude that by the fourteenth century they were already being used for houses at all levels of society and for all building types. He also found that slaters ran their own quarries, a practice which still continues in some quarrying centres today (Moorhouse 1990).

From the seventeenth century the demand for more buildings using more substantial materials encouraged the development of the stone industry and the manufacture of stone slates. At that time stone slate delves were still largely supplying a small local market. This was to change during the Industrial Revolution when the increasing demand for flagging for pavements and factory floors and roofing for houses and mills stimulated the delving of fissile rocks. This expansion was a feature of both rural and industrial regions but it was in the Carboniferous rocks of Lancashire and Yorkshire, and the Devonian of Angus and Caithness in which the scale of operations could truly be called industrial. Thousands of tons of flagging were exported to all parts of Britain and even to North America. Roofing was also produced in large quantities but tended to remain close to the centres of production. This was a reflection of the by now well-established segmen-

Table 2. Market size for stone-slates and the supply situation in England.

Market		Size	Supply within the region
Dorset and Somerset	Purbeck	modest	inadequate
	Mainland	modest	virtually none
Cotswolds		large	limited but improving
East Midlands		modest	inadequate
South East		modest	none at present
Welsh Marches and Bristol		complex, modest	inadequate or none
Derbyshire and Peak Park		large	inadequate but may improve soon
Lancashire and West Yorkshire		large	improving but threatened by imports
North Yorkshire	Carboniferous Jurassic	large small	inadequate none
Durham and Northumberland		modest	inadequate and deteriorating
Cumbria	Permian Carboniferous	modest modest	none inadequate

tation of the market for roofing materials, essentially a segmentation based on price, as the cost of moving the heavy stone slates limited their geographical use.

Nonetheless stone was still a cheap form of roofing close to the delves. But this was about to change. Thinner metamorphic slates, lighter and therefore cheaper to transport, were about to take a major segment of the roofing market. At the forefront of this development were the two Welsh slate production centres between Bangor and the Nantlle Valley in the west of Snowdonia and around Blaenau Ffestiniog to the east. During the nineteenth century they became huge concerns employing thousands of workers. Their products were initially transported by sea around the coast of Britain and later via the canal and railroad systems. The railway link from Chester to Holyhead was opened in 1848 allowing the western quarries to gain access to all parts of Britain. By 1850 Penrhyn Quarry near Bangor was employing nearly 2000 men. The Cambrian Coast railway line followed in 1867 and the main London and North Western line reached Blaenau Ffestiniog in 1879. Eventually the Welsh quarries were to almost supplant the use of stone slates. But they did not die out completely. Production continued but under increasingly difficult circumstances. The loss of workers during two world wars and factory work and the better wages and conditions it offered, took their toll and by the 1950s in many stone-quarrying areas roofing production was only clinging on as the result of local building conservation controls and planning policies which supported the use of stone slates on council housing. Competition was now also coming from industrial roof products: concrete tiles and, later, artificial slates. The period from 1950 to 1980 saw the decline continue with more and more stone slate delves closing down.

From the 1980s there was a renaissance in the use of natural materials for building, including roofing. The Welsh slate industry had gone through a period of rationalisation into a few large concerns and was able to respond to the increased demand, but the small stone slate delves were not and as a result the attempts by building owners, architects and conservation bodies to save stone roofs were largely frustrated by inadequate and unreliable production. It was this sense of frustration which prompted Derbyshire County Council and the Peak District National Park to call a public meeting in 1993 and which eventually led to the English Heritage *Roofs of England* campaign to regenerate the industry, and to this study of the state of stone roofing in England.

Markets and supply

The markets for stone slates and the current supply situation are summarised in Table 2. Since the beginning of the *Roofs of England* campaign the situation has improved, especially in the two largest markets, Yorkshire/Lancashire and the Cotswolds. In Derbyshire, thanks to sustained local efforts, supplies will probably improve in the near future, and in Horsham new stone slates may soon be available for the first time in 60 years. Pennant Sandstone, used widely in Bristol, Gloucestershire and South Wales, is at last available again from a new delph near Pontardawe, and delves working slates from the Forest Marble (Somerset and the Cotswolds), Fullers Earth rock (Cotswolds) (Fig 2), Millstone Grit and Coal

Figure 2. It is now possible to select some appropriate regional types of stone slate in the Cotswolds. Fullers Earth, Eyford Member stone from Grange Hill Quarry (© Terry Hughes).

Figure 3. New Harnage Slates on St Michael's and All Angels Church, Pitchford, Shropshire (© Terry Hughes).

Measures (Lancashire and Yorkshire) are in production. The Harnage delph at Acton, near Shrewsbury, was reopened temporarily during 1999, originally to renew the roof of St Michael's and All Angel's church with the very distinctive Harnage slate (Fig 3), but has subsequently supplied three other important historic buildings. It is available for further working whenever another Harnage roof is needed (see Wood and Hughes, this volume).

Most of the stone slate types are very distinctive and contribute to the unique character of their area. However, within these markets, there are individual slates which may not be distinctive enough to justify re-establishing continuous production and consequently the use of substitutes of similar appearance may be acceptable. But care is needed when selecting a substitute. Appearance and geological attributes are both important but it could be argued that it would be more appropriate to select the geographically closest alternative on the basis that this is what would have been used had the original stone slate not existed.

Manufacture

The size of the market for stone slates varies enormously. Of course, the size depends on how specific the definition of a slate type is, but some types are so distinctive that they must be acknowledged as examples worth conserving even if there are only very few remaining roofs. The Permian sandstone of the Eden Valley in Cumbria and Harnage slate are examples. For types with too small a market to sustain continuing production a decision has to be taken about how best to conserve the roofs. There are only a few sustainable options: building-specific or intermittent production from the original delph, or substitution by the 'same' stone slate from another source.

- **Building-specific** or **one-off production** to conserve an individual roof is an attractive option because it will provide an authentic replacement with all the subtleties of colour, texture, size mix etc which are so important to local distinctiveness, at least as far as the geological vagaries of delves will allow. This option has been successfully attempted for Harnage slates.
- **Intermittent production**: if it is known that there will be a continuing but infrequent need for a specific stone slate type it is possible to obtain minerals planning consent to work a delph at appropriate intervals. This is unlikely to lead to planning problems if there are changes of land ownership as the permission is granted on the land and not to the owners. But difficulties could arise over the production and economics if there is no continuity in the delving and manufacturing. For this reason it is preferable that a contract be established with a company which would produce sufficient slates as and when required. There will be benefits of scale, and a reduction in product cost, if the market can be managed so that relatively large amounts are produced infrequently. The organisations with influence over the rate of roofing renovations can have a useful role in this respect. The National Trust and local authorities with substantial building holdings should be encouraged to schedule renovations in co-operation with manufacturers. Similarly Conservation Area Partnership Schemes and similar initiatives should include a review of product availability to ensure that the greatest benefits of scale are secured and supply difficulties avoided. This approach may have to include an element of stock-piling for roofing work in later years.
- **Small-scale continuous production**: several examples of this exist, sometimes only involving a workforce of one or two people. When only a few people are involved it is both a strength and a weakness. The closer the link between production and income, the stronger the motivation and commitment to produce the goods. On the other hand, the 'one-man-band' operation is extremely vulnerable to sickness and injury. During 1997 the total output of one stone slate type was halted following an accident and the re-roofing of several buildings was completely disrupted, even to the extent of some stone slates being replaced by other, inappropriate, slates or clay tiles. Also, if a stone slate type is only available from a single source there will be no pressure to hold down the price, with the risk that less roofs will be conserved.
- **Full-scale production**: several markets will sustain full-scale production and some areas, including, for example, the Cotswolds (Figs 4–10), and Yorkshire, are large enough for several companies to operate in competition with each other with the potential for reductions in product cost. In this situation vigilance will be needed to ensure that downward pressure on the selling price does not result in undesirable production techniques, resulting in a degradation of the product or of substitution by inappropriate slate types. Substantial commercial operations of this type may have the capacity to operate other delves for building-specific or intermittent production of other stone slate types in other regions.
- **Substitution by stone slates from another source**: by force of circumstances this option has been adopted in some regions for many years. Unfortunately it has

Figure 4. Stone slate delving is usually shallow and on a small scale. The importance of keeping the slates as large as possible encourages hand methods. The Cotswold Stone Tile Company delph in the Forest Marble near Crudwell, Wiltshire, is shallow, extending to only 4m depth (13–14 ft) and currently only three-quarters of an acre (one-third of a hectare) is being worked. (The skyline in the photograph shows a 3.5m tall embankment.) It is expected that the current working area will be reinstated and returned to agriculture within 12 to 18 months (© Cotswold Stone Tile Company).

Figure 5. In the Forest Marble stone slates often 'float' in a layer of clay. Here they are being removed by hand. In contrast, in sandstone quarries the slates are levered out of 'solid' rock (© Cotswold Stone Tile Company).

been applied in an unstructured way, with architects and contractors forced to make choices as best they could. The result has been that often inappropriate stone slates have been selected as replacements. In the Brandsby area of North Yorkshire, for example, laudable attempts to conserve the stone roofs have resulted in the use of Carboniferous sandstones on erstwhile limestone roofs, with a radical change of appearance. A more suitable stone slate could have been obtained from the Cotswolds region, always assuming that the Cotswold supply situation was satisfactory. This has seldom been the case in recent years and that in itself led to the importation of limestone slates from France.

Figure 6. Forest Marble slates as received from the delph (© Cotswold Stone Tile Company).

Figure 7. Sawing the rough slates to size and shape (© Cotswold Stone Tile Company).

There are a number of factors which must be taken into account when choosing to source slates from outside a region.

- The **appearance** of the slate should be similar to the local stone slate, both within a short time of being installed and after long-term weathering. Anyone wanting to select a stone slate from another region in the UK could research this reasonably easily and only the more extreme climatic factors need be taken into account. That said, it would be a lot easier to do this if stone slate regions had databases of local roofing stones, ideally with colour photographs of new and weathered slates. For sources outside the UK this is a more difficult exercise and climatic effects are likely to have a more significant effect on appearance.
- **Format**: substitute slates must be available in a *similar* size range, thickness and edge treatment etc. Stone slates are beginning to be imported from abroad, to fill the gap in the local supply and because they are cheaper than local products. They may eventually

Figure 8. Hand-dressing the slate's edges. The form of the edge can be tailored to suit the local tradition (© Cotswold Stone Tile Company).

capture the market. In an effort to reduce costs and to simplify their inventory, importers are stocking a reduced range of thicknesses, sizes and stone slate types and this is leading to less individuality in the roofscape of, for example, West Yorkshire.

- **Strength and durability**: generally stone slates are so thick that they will resist all the normal loadings they will experience on the roof, but they should also provide a strong peg or nail hole and not be too brittle to withstand manufacturing and site handling. They should be of similar durability to the local stone. In the absence of performance tests this is difficult to assess because many stone slates have a mineralogy which may be theoretically suspect but can be satisfactory in practice. On the other hand, stone slates may have performed well in their own region but would not do so in a different climate or under a different method of roof construction. More worryingly, stone from abroad is being imported which has never been used for roofing, let alone in a British climate. At the very least imported stone slates with no track record should be assessed geologically and advice sought on appropriate testing methods. A guide to testing stone slates is in preparation for the Building Research Establishment.
- **Cost**: ideally substitutes should be cheaper than local production or have some other advantage such as reduced environmental and social impact. There are immense problems with sourcing stone imports from countries where environmental and health and safety controls are poor. It should not be overlooked that the cost of products manufactured in the UK is often high for the good reason that these controls are stringent here. Importing stone from outside the UK may in practice be an export of environmental and health impacts.
- **Impact on local production**: once an external substitute is 'approved' it will reduce the viability of a local industry. Cost advantages of imports may be short-lived. Selling price is always a marketing decision and once a market has been penetrated on the back of cheap products and the local competition eliminated, the price may well increase.

Figure 9. Drilling the peg/nail holes (© Cotswold Stone Tile Company).

Figure 10. Sorting the finished slates by length ready for making up into an order (© Cotswold Stone Tile Company).

New delves

A number of delves are being opened or reopened, sometimes without any formal assessment of the rock's suitability. Durability tests which relate specifically to roof exposure criteria are needed. In the interim, the best evidence of suitability is the history of the stone slates from the specific delph. Failing that, the advice of a geologist should be sought.

Experience over the last five years has shown that manufacturing initiatives will be more likely to succeed where there is a strong commercial involvement from the beginning. People with the required entrepreneurial attributes are relatively easily found in industrial and commercial regions but less so in the rural areas. However, farmers have all the required attributes and the production of stone slates and associated products can sit quite happily alongside farming activities. A number of successful delves are now being operated by farmers.

The need for quality control

The conservation of stone roofs and the manufacture of stone slates is subject to the same commercial and legislative influences as any other sector of the construction industry. These and other developments have begun to affect the production, availability and use of stone slates. There is a strong conservation argument for maintaining traditional production methods to conserve the appearance of the slates. However if the cost of production becomes so high that more roofs are lost as a result, then overall building conservation objectives will not be achieved.

The single most important change which would improve the level of stone roof conservation would be a significant drop in the price of stone slates. Such a drop can only be achieved by an increase in the scale of production, a reduction in manufacturing costs (and the former will lead to the latter) or imports from countries with lower labour costs and overheads.

The manufacture of stone slates has a very high labour content and there is therefore considerable potential for productivity gains through mechanisation. The risk is that the visual quality of the product may be degraded as a consequence and there are several instances where this has been the case to a completely unacceptable extent.

In an effort to improve supply and reduce costs, several manufacturers are using rotary saws to cut slates to size or thickness. This could have technical and aesthetic consequences. The use of inappropriate rock or imprecise sawing could lead to slates with unsatisfactory durability and the appearance of the slates could be non-natural, which in turn will alter the appearance of the roofs. Reduced durability is entirely unacceptable, but it has been argued that if a slightly degraded appearance made the slate cheaper to buy, then it might help conservation objectives by saving more roofs. Nonetheless, edges must be dressed after sawing if the appearance of historic roofs is to be conserved.

What is needed is a co-operative approach between the manufacturers and the conservation bodies to decide what is acceptable. In the end if the conservation of buildings can only be achieved through the use of new techniques such as sawing to size or thickness and subsequent tooling to produce an acceptable appearance, this may be a necessary sacrifice. Perhaps what is needed is a structured approach whereby the buildings of greatest importance are conserved to the highest specification using only authentic and traditional manufacturing and building methods, and a more relaxed approach to buildings of less importance.

Some manufacturers offer a bespoke mix of slate sizes to fit roof slope dimensions or to match existing gauges for individual roofs. This technique is very attractive to roofers because it makes their job easier and eliminates some of the cost uncertainties. But it risks losing the natural variety of sizes and mixes of sizes from roof to roof which is so important an aspect of local distinctiveness. It should only be used with care and an understanding of the implications. There is also a distinct risk that if stone slates are supplied in the *exact* quantities to cover each course there will be insufficient choice of widths to achieve a satisfactory sidelap. This could have very serious consequences for the water-tightness of the roof.

At present, the basis on which stone slates are approved for use on particular buildings, in conservation schemes or within regions, is largely unstructured and *ad hoc*. Sometimes regional types are assumed to have specific characteristics when in fact there are a range of variants. Sometimes wholly inappropriate types are selected for convenience or out of ignorance of the availability of more suitable slates. Of course, just as often there are no options on what is used simply because there is only one source of supply. Nonetheless there is a need for structures within which informed decisions can be taken. The South Pennine Grey Slates study developed a type list for that region (Figs 11–18, Table 3) and this could be usefully extended for other regions. It should also include details of other characteristics such as size ranges, edge treatments etc.

Minerals planning

In none of the regions researched was the mineral planning policy found to present obstacles to the opening or re-opening of delves to produce stone slates. On the contrary, the policies are supportive and only require a need to be demonstrated and the normal impact controls to be applied, to gain the active help of the mineral planning officers.

On the other hand, the process of applying for planning permission is an obstacle in so far as it is comparatively expensive for individuals wishing to set up a small operation, and also difficult to understand. A guide to making a mineral planning application for stone slate quarries, written in simple language, has been produced by the Stone Roofing Working Group. [5]

Environmental controls

Environmental controls on delving are becoming ever more stringent. Desirable as the protection of the natural environment is, if the controls are implemented without due regard to the importance of stone to the built heritage they will pose a serious threat to building conservation. When a specific stone is required to conserve a building, a number of factors have to coincide if it is to be delved. There must be an economically viable source, adequate finance, an existing company (or, more commonly, an entrepreneur) willing to operate a delph and then planning permission must be granted. These factors coincide so infrequently for most stone slate production ventures that a planning application should receive support rather than obstruction.

Figure 11. Derbyshire and Peak Park stone slates. Yorkstone type: flat, featureless, without substantial stepped bedding, fine- to medium-grained, buff to dark brown. Munsell 10YR (© Terry Hughes).

Figure 12. Derbyshire and Peak Park stone slates. Kerridge type: flat, featureless, without stepped bedding, fine-grained, grey mica surface. Mica on the surface will quickly flake off once the roof has been built (© Terry Hughes).

Figure 13. Derbyshire and Peak Park stone slates. Cracken Edge type: textured, with or without stepped bedding, fine- to coarse-grained, white and buff to dark brown (© Oxford Brookes University).

Figure 14. Derbyshire and Peak Park stone slates. Teggs Nose type: textured, with or without stepped bedding, fine- to medium-grained, pink (© Oxford Brookes University).

Figure 15. Derbyshire and Peak Park stone slates. Freebirch type: strongly textured or ripple etc marked, fine- to medium-grained, buff to dark brown and olive to grey showing tooling to produce a flat surface (© Oxford Brookes University).

Figure 16. Derbyshire and Peak Park stone slates. Wirksworth type: strongly textured, fine- to medium-grained, pink to red. Munsell 10R, scale in centimetres (© Terry Hughes).

Figure 17. Derbyshire and Peak Park stone slates. Magnesian Limestone type: a dolostone from Hardwick Hall in east Derbyshire (© Oxford Brookes University).

Figure 18. Derbyshire and Peak Park stone slates. Magnesian Limestone type showing the weathered colour (© Terry Hughes).

Figure 19. The old delph at Reeve Edge in the Peak District National Park. Although a major source of roofing for about 100 years, the delph is very small (scale shown by the sheep) (© Terry Hughes).

Table 3. The generic slate types proposed for the South Pennines (Hughes 1996 vol 2, 7).

Generic type	Description
Yorkstone	flat, featureless, without substantial stepped bedding, fine to medium grained, buff to dark brown
Kerridge	flat, featureless, without stepped bedding, fine grained, grey mica surface
Cracken Edge	textured, with or without stepped bedding, fine to coarse grained, white and buff to dark brown
Teggs Nose	textured, with or without stepped bedding, fine to medium grained, pink
Freebirch	strongly textured or ripple etc marked, fine to medium grained, buff to dark brown and olive to grey
Wirksworth	strongly textured, fine to medium grained, pink to red
Whitwell	strongly textured or ripple etc marked, fine grained, grey or pink, magnesian limestone

The fact is that stone slate delves are always very small with very slight impacts on the environment (Fig 19). But if planning permission is refused because environmental controls take precedence over all other considerations then the built environment will never be properly conserved. This problem is particularly acute within National Parks and similarly designated areas where there is a distinct danger that legislation controlling a Park's responsibility to conserve the natural heritage will prevent them executing their responsibility to conserve the built heritage which does not have such a stringent legislative background. Additionally, global conservation objectives will not be achieved by preventing production of stone slates in a small, local delph only to approve their importation from a country thousands of miles away with all the attendant waste of fuel.

Conclusions

Decisions to move away from traditional local stone slates to alternatives from other regions, other countries or new quarries often have to be made without a satisfactory basis for comparison.

At present there is little interchange between manufacturers and conservation authorities about the suitability of novel production techniques. This results in stone slates with entirely unsuitable features being used to the visual, and sometimes the technical, detriment of the building.

Work is required in three areas to provide decision-makers with a structure in which to work:

- a definition of the visual characteristics for each stone slate type
- an agreed basis for what is acceptable as a deviation from the traditional characteristics consequent on novel production techniques and the buildings on which these deviations are acceptable
- tests for the durability of existing products so that they can be used outside their historic areas and for slates obtained from old delves so that only suitable delves are reactivated.

Individuals with limited financial resources need access to help in making mineral planning applications.

Roof construction

Traditional construction

This is a surprisingly under-researched topic. At best, roof renovation is done on an 'as is' basis without recording the details of the existing structure. Without accurately researched knowledge and in an environment where there is little or no formal training in stone roofing design and construction, there is a risk that bad practice and faulty construction will be perpetuated. In 1998 an English Heritage Stone Slate Technical Advice Note was published, written as a generic guide and, in part, intended to encourage regional and local organisations to publish their own detailed guides to local stone roof construction. So far this has largely not happened and this serious gap in the technical basis for the specification and construction of these roofs continues to hamper their conservation.

To investigate the construction of a stone slate roof correctly it should be carefully stripped and recorded by someone who understands how slating works (Fig 20). Simply photographing the exterior of the roof will provide almost no useful information about the constructional details.

Figure 20. Recording the construction of the roof at Lady Margaret Hungerford Almshouses, Corsham, Wiltshire, prior to re-roofing (© Terry Hughes).

Modern roof construction techniques

Modern methods of constructing roofs are adopted almost without demur in some regions, but are entirely rejected in others, for the best conservation motives. There is considerable potential for errors in applying new methods, especially with insulation and ventilation, which is a rapidly evolving field, largely driven by product manufacturers.

Roof conservation

In spite of conservation legislation and grant support, stone roofs continue to be lost (Fig 21). The reasons are the unavailability of new stone slates (which is being addressed), the limited extent of conservation control, the cost of re-roofing, and insufficient grants to offset the extra cost of the handmade stone slates and the cost of fixing them.

Unsurprisingly, roofs on unlisted buildings are continually lost. This situation is aggravated by the ready market for reclaimed slates which encourages their removal. Some conservation bodies actively support the use of stone slates from one building, or group of buildings, on another. This is quite simply not sustainable. It will inevitably undermine the viability of the manufacture of new slates and without new stone slates to feed into the re-roofing cycle, it will ultimately result in the disappearance of all stone roofs.

The cost of stone slates and the expense of re-roofing work is fundamental to how many roofs will be lost. There is some potential for the price of new slates to be reduced by adopting novel manufacturing methods, albeit perhaps to the slight detriment of their appearance, but unless more money is available to support their use in conservation, roofs will continue to disappear. The point may already have been reached where the grant-giving bodies may be forced to narrow the scope of buildings which they support so that the available money can be put to the best use.

The *Roofs of England* initiative to try to reinstate the production of stone slates marked a fundamental change in the approach to conserving buildings. Rather than simply putting money and effort directly into the conservation of individual buildings it attempted to solve the problem of the impossibility of conserving buildings without the authentic materials from which they were constructed. Unfortunately, all other sources of grants within the building conservation field can only be applied to the building, not to establishing the manufacture of the stone slates where they are not currently available. This gap in funding is seriously hampering the opening of delves which are not economically viable because of their small markets.

Figure 21. Stone roofs continue to be lost to other products, shown by these Welsh slates in a gritstone terrace in Tideswell, Derbyshire (© Terry Hughes).

Training

Currently there is little formal training available for designers and conservation professionals. The short courses which are run from time to time hardly scratch the surface of what needs to be understood and often only attract a few students. This does not make the best use of the trainers' time: fewer larger courses would be more time-efficient and more economically viable.

The situation for slaters is even worse, with virtually no formal training in stone-roofing techniques available. Most slaters rely on picking up on-the-job training from other roofers. This risks the perpetuation of bad practice. There is virtually no training available in historic roof conservation as distinct from roof construction.

Figure 22. It is difficult to re-establish delving of stone slates which only have a small market. Those like the red Permian sandstones of the Eden Valley are distinctive and important, but without determined support and financial help they are likely to disappear entirely (© Terry Hughes).

Action points

- There is an urgent need for the construction of older stone roofs to be thoroughly researched before the remaining examples, which date from the time when the techniques were well understood, are renewed.
- The conservation bodies need to decide within their region which stone slates should be manufactured because they are important and distinctive, and which could be substituted from other sources. This will provide the basis for re-establishing regional stone slate manufacture.
- A logical protocol is needed by which informed decisions on the selection of substitutes can be made. For each region this protocol requires, at the least, a detailed description, including pictures, of each of the stone slate types and their localities.
- Durability tests for stone slates would ensure that unsuitable products are not manufactured in the UK or brought in from abroad.
- The conservation bodies need to debate the costs and benefits of technological initiatives and to set limits on what is acceptable.
- The adoption of new manufacturing technologies may be creating a new 'vernacular'. This may be acceptable for new buildings but not for the conservation of existing roofs. Similarly, modern roof construction practice is changing roofs. Practitioners need to seek advice on whether and how best to adopt modern roof construction products and techniques. It may be that a more precise statement of conservation policy in this respect needs to be articulated and disseminated. At the least, guidance is needed on the issues which need to be considered so that the appropriate decisions can be taken without everyone having to go through the whole process from the beginning each time.
- There is an urgent need to direct financial support to the fundamental problem of winning new slate in areas where the market is small (Fig 22).

To address some of the issues listed above the English Heritage Stone Roof Working Group was set up in 1998. It has produced a guide to help potential delph owners through the intricacies of mineral planning applications and is currently writing a best practice guide for conserving stone roofs.[5] It is also co-ordinating and, in some cases, writing local guides for specific roof types. These are based on a framework which covers the issues that need to be addressed in such guides and which is available to other authors. The members are also discussing the issues of minerals and natural environment planning controls with the Department of the Environment Transport and the Regions and English Nature and hope, in the future, to establish stone-roof courses for designers and slaters.

DORSET AND SOMERSET

For convenience and to avoid repetition the discussions of the counties of Somerset and Dorset have been combined.

Figure 23. Montacute, Somerset. The Upper Lias Ham Hill stone produced roofing in the past, but prolonged unavailability has resulted in the progressive change of the roof covering to other products: usually metamorphic slates (© Terry Hughes).

Table 4. Sources and historical stone slate production centres of Jurassic and Cretaceous age in Somerset and Dorset.

Age		Rock unit	Production locations
Cretaceous	Durlston Formation (Middle and Upper 'Purbeck Beds')		Vale of Wardour, Isles of Purbeck and Portland
Upper Jurassic	Lulworth Formation (Lower 'Purbeck Beds')		Isles of Purbeck and Portland
Upper Jurassic	Portland Beds Ampthill and Kimmeridge Corallian Upper Oxford Clay		
Middle Jurassic	Middle and Lower Oxford Clay		
Middle Jurassic	Kellaways Beds		
Middle Jurassic	Great Oolite	Cornbrash	
Middle Jurassic	Great Oolite	Forest Marble	Fleet, Bothenhampton, Yetminster, Lillington, Longburton, Yenston, Stalbridge
Middle Jurassic	Great Oolite	Upper Fuller's Earth Clay	
Middle Jurassic	Great Oolite	Fuller's Earth Rock	
Middle Jurassic	Great Oolite	Lower Fuller's Earth Clay	
Middle Jurassic	Great Oolite	Crackment Limestone	
Lower Jurassic	Inferior Oolite		
Lower Jurassic	Bridport Sand, upper part		
Lower Jurassic	Upper Lias		Ham Hill
Lower Jurassic	Middle Lias		
Lower Jurassic	Lower Lias		Queen Camel, Charlton Mackrell area

Table 5. Section through the old source of roofing at Ham Hill.

Beds	Thickness	Uses
Weathered limestone	0.8 m (31.5 in)	
Thinly bedded and flaggy limestone	8.9 m (29 feet)	Tilestone source
Alternating thin limestone and sand	1.3 m (4.25 feet)	
Thin sandstones, calcareous sandstones and limestone	4.5 m (14.75 feet)	
Massive limestone	12.5 m (41 feet)	Main building stone

Geology of Dorset and Somerset

All the traditional stone slates of this region have been obtained from the Jurassic and the Cretaceous although there is some evidence that Pennant Sandstone from Bristol has been used occasionally in recent years to repair limestone roofs. The Jurassic and Cretaceous locations are shown in Table 4.

Lower Lias

Woodward (1893, 295) recorded that the Lower Lias limestones yielded thin stone tiles used for roofing at Queen Camel near Sparkford but gave no further information. It has also been suggested that the Lias was worked for roofing in the Charlton Mackeral and Charlton Adam area. From the lower part of the Blue Lias, roofing has been found in excavations at Roman sites, for example at Lopen, east of Ilminster (Hugh Pruddon, personal communication).

No stone roofs were found during a brief survey around the Charltons and at Queen Camel. This might be because they had a short life and have subsequently been replaced by tiles, as would be the case with, for example, some Blue Lias which can be split thinly. The White and Blue Lias 'vary greatly in durability depending on their stratigraphical position, the flaggy limestone of the Blue Lias (widely used for paving) is hard and resistant to weathering; the overlying, nodular Blue Lias limestones are generally less satisfactory' (Kellaway and Welch 1993, 157).

Upper Lias

At Ham Hill, roofing stone (Fig 23) was obtained from quarries which have existed from at least Saxon times. This is now a protected site. The limestone beds near the surface had been rendered fissile by peri-glacial action (Table 5).

> The ground from which the stone for all our old buildings was obtained is on the western side of the hill, and mostly in the parish of Norton. These old workings were only about twenty feet deep in stone, at the most, and the heading was of rubble and thin layers of stone. The stone

Figure 24. Forest Marble slates quarried near Yetminster, Dorset (© Terry Hughes).

> tiles, with which so many of our buildings are covered were quarried near the surface, over the workable stone, chiefly from the north part of the hill. Instead of the ochre or sand beds of the deep modern quarries, there were here thin layers of hard stone, which were worked to an even thickness by a tile-pick. The working of tiles is now a lost art on the hill. (Traske 1898, 217)

South of the Saxon quarry, there are two faults which have downthrown the roofing beds. They are now found near the bottom of the present quarry where they are too compact to be suitable for splitting (David Jefferson, personal communication).

It appears that fissile rock is available near the surface in the area (in the Ham Hill Stone Co quarry) but only in small quantities. The potential to work these will be dependent on the commercial plans of the quarry which are likely to concentrate on masonry stone from deeper levels. Nonetheless, there is an opportunity for either intermittent delving for specific buildings, or for a specialist stone-roofing manufacturer to operate these beds.

Figure 25. Forest Marble on a church porch at Lillington, Dorset. Characteristically, Forest Marble roofs are heavily textured because of the the inherent roughness and twist of the slates (© Terry Hughes).

It is also relevant that Ham Hill is a Scheduled Ancient Monument.

In the absence of slates from naturally fissile rock, for whatever reason, there is the option of producing roofing sawn from masonry block. At the time of writing this was being actively pursued and the edge and surface dressing which was traditional for this stone was looking very promising. Subject to durability testing, it may be that slates with an authentic appearance will be possible.

Ham Hill stone slates have been used over quite a large area, for example as far south as Mapperton House near Beaminster, but it is not known whether they are an original feature or later repairs.

Forest Marble

Outside Purbeck and Portland, the main stone used for roofing in Dorset has been the Forest Marble. It occurs intermittently along the coast between Portland and Bridport and then northwards to near Crewkerne. From there, it turns west towards Sherborne and then northwards, passing to the east of Frome and into the Cotswolds.

- **Portland to Bridport**. Delves which supplied the area around Fleet and Langton Herring produced thick, ripple-textured stone slates and at Fleet House Farm large pieces of stone suitable for roofing are turned up during ploughing. At Bothenhampton, shelly limestones were used for roofing but sandstone roofing may also have been available from the delves to the south of the town (Jo Thomas, personal communication).
- **Sherborne**. Stone slates have been produced within the immediate vicinity of Sherborne, especially from the ridge to the south of the town from Yetminster through Lillington (Figs 24 and 25) and Longburton to Yenston. This area is close enough to Queen Camel and even Ham Hill for some buildings to have been roofed with Lias stone.
- **Stalbridge**. At Stalbridge fissile stone has been delved from the Forest Marble Formation, producing unevenly surfaced, comparatively thick stone slates and distinctive roofs (Fig 26).

Figure 26. At Stalbridge the Forest Marble produces particularly heavy and uneven stone slates (© Terry Hughes).

Corallian

> Limestone from the Corallian beds has provided a source of building stone for centuries. The older parts of Marnhull, Hinton St Mary and Sturminster Newton, and many of the other villages, are built of the stone. The principal limestones worked were the Todber Freestone, the Clavellata Beds and the Cucklington Oolite. White noted that blue shelly limestone (Clavellata Beds and Cucklington Oolite) was used for building and stout flagstones for flooring and drystone walling. (Bristow *et al* 1995, 154)

Although the Corallian has been used as a building stone and for flagstones locally and for roofing in the Cotswold region, it has not been possible to confirm its use as roofing is this area. Several of the villages named above and Gillingham have buildings with stone roofs which may have come from the thinner, flaggy beds in the Corallian but could equally have been obtained from the nearby Forest Marble at Stalbridge and near Sherborne, or the Purbeck Beds from Tisbury in the Vale of Wardour. Dobson includes a photograph of a 'Stone slate quarried near Gillingham, Dorset, from a roof laid in the eleventh century', unfortunately without stating how near the quarry was (Dobson 1960, 15).

The Isles of Purbeck and Portland

> In the Isle of Purbeck and in the vicinity of Sherborne many houses are covered with stone tiles which are flat stones often an inch thick and require very strong timbers to support them but the major part of the buildings in the villages and in some of the towns are covered with what is called reed thatch or wheat straw which has not been bruised by the flail. (Stevenson 1812, 86–7)

> Purbeck Stone. The quarries, shores, and cliffs, on the south side of Purbeck, afford an inexhaustible fund of natural curiosity ... The quarries are chiefly near Kingston, Worth, Langton, Swannage. In many parts of the island is a stone that rises thin, and is used for tiling; also a hard paving stone that sweats against a change of weather. (Stevenson 1812, 55)

Figure 27. Purbeck slates waiting to be reused at Langton Matravers. In some of the slates the peg hole is several inches from the top edge. This is because they were used in a course of shorter slates rather than because a higher hole would be too weak. Lowering the peg hole keeps the tail of the slates in a straight line. On courses of long slates there is ample room for the extra length above the hole to sit between the battens (© Terry Hughes).

The Isles of Purbeck and Portland have been quarrying centres since at least Roman times and both contain fissile rock suitable for roofing (Fig 27). The stone slate sources are found in the Purbeck Beds. [6] Originally judged to be located at the top of the Jurassic, they are now considered to straddle the Jurassic/Cretaceous boundary. The base of the Cretaceous is taken at the base of the Cinder Bed, which lies within the Middle Purbeck Beds. That part of the Purbeck Beds lying within the Jurassic is now termed the Lulworth Formation, and the portion of the Purbeck Beds now considered to be Cretaceous in age have been named the Durlston Formation. The names of the beds and their stratigraphic relationships are complex because they are not consistently present throughout the area. Consequently, it is difficult to align old records with the modern names of beds. Table 6 provides an approximate alignment of an early geological record (Webster 1826) with the names in current use (Saville 1986). The Upper Purbeck beds are shown in Table 7.

The Middle Purbeck Beds on the Isle of Purbeck are the best-known source of roofing (Fig 28). The use of stone slates only became common during the eighteenth century as a by-product of paving production. During that period the Downs Vein was the preferred and most productive bed and has the best reputation for durability on roofs, but the Wetson, Thornback, Roach, and Grub veins also split thin enough for roofing use from time to time (Fig 29).

There is historical and colloquial evidence that the Lower Purbeck Slatt beds on the Isle of Portland were fissile enough for roofing, and that they were widely distributed by sea (Gerald Emerton, personal communication). One roofing source was the slatt beds in the thin, fine-grained limestones of the northern quarries (Jo

Table 6. Approximate alignment of historical and modern 'Purbeck Bed' names.

Formation	Age	DB*		Webster's Beds (1826)	Modern name (Saville 1986)		
	Upper				See Table 7		
Cretaceous, Durlston Formation	Middle	140	19	Royal Shiver etc		Royal Rag	
		139	20	Devils bed Shiver etc			
			21	Iron bed Shiver etc			
		133	22	Red rag Shiver etc			
			23	Upper rag Shiver etc			
			24	Under rag Shiver etc			
		129	25	Lead bed		Mead	
			26	Shid bed			
			27	Shingle, 2 beds Shiver etc			
		126	28	Grub Shiver etc		Grub	
		125	29	Roach in 4 beds, good	Freestone Vein	Roach	
			30	Grey bed, good			
		125	31	Thornback, good		Thornback	
						Wetson	
			32	Freestone, good Shiver etc		Freestone	
		121	33	Lias, not worked			
		116	34	Lias rag			
		114	35	Downs vein rag Shiver etc		Laper Downs Vein	
			36	Grey bed, No 2, not worked		Grey Bed	
			37	Hone Hone, not worked marly, Shiver			
			38	Grey bed, not worked			
		113	39 40	Upper bed } Lower bed } these divide into slates Shiver	Downs Vein		Pon Mangy White Pon & Underset Clearall Pon Bottom Under Bottom
			41	Cinder, useless, a bed of oysters. Stone, not worked		Cinder	
Jurassic, Lulworth Formation	Lower		42	Button; splits into slates		Button	
			43	Feather, good		Feather	
			44	Cap, used only in backing		Cap	
			45	Flint, used only in backing Shiver			
						Sky Bed	
			46	Upper, five bed, sometimes splits into 5		Five or Six Bed	
			47	Under five-bed, sometimes splits into 5	New Vein		
			48	White bed		White Bed	
						Brassy Bed	
			49	Tomb-stone, good		Thompson Bed	
			50	Pudding, inferior			
			51	Sheer, used in backing Shiver			
			52	Flint, used only in backing			
Portland Stone				Shrimp Bed at Purbeck, Roach at Portland			

* A number of schemes exist for identifying beds in the region. The one for this locality is based on exposures at Durlston Bay hence the DB numbrs. The allocation of DB numbers to webster's beds is based on West.

Figure 28. Corfe Castle, a quintessential stone slate village (© Terry Hughes).

Thomas, personal communication), but it is not known whether all these beds were equally suitable. The Thin Slatt, for example, might imply that they were light enough to be used as slates or too weak to be suitable. They seem to have largely disappeared from the buildings on the island, although the roof of a cottage and attached store was repaired with local stone within recent years (Peter Trim, personal communication).

Table 7. The Upper 'Purbeck Beds' listed by Webster.

Bed No.	Bed name	
1	Marble, not worked at present	Purbeck Marble
2	Marble rag, Shiver, marl etc	
3	Single leaper, good stone	
	Shiver, marl etc	
4	Step bed, good stone	
5	Grey rag, good stone	
	Shiver, marl etc	
6	Toad's eye, a very hard but good stone	
	Shiver	
7	Good bed, bad stone	
	Shiver, etc	
8	Pitching-stone bed, good stone	
9	Tomb-stone bed, good stone	
10	Yellow bed, bad, not worked	
11	White roach, not very good	
	White earth, a marle	
12	Leaper, excellent	Laper
13	White bed, excellent	
14	Soft bed, excellent	Lane-end Vein or Laning-end Vein
15	Hard bed, excellent	
16	Mock hard bed	
	Earth	
17	Pitcher bed	
18	Backing	
	Shiver etc	

Within the Lower Purbeck several rocks or beds are named Slatt, and Damon gives the following description of one:

> Bed 2 Slate. The deposit constituting the upper stratum of the Isle of Portland though in some quarries 10 feet in thickness, varies in other parts of the Island from less than 3 feet to 15 feet. It is chiefly made up of a hard stone shivered [7] into layers of about an inch thick. The whole bed is much shattered, and the stratification most irregular. (Damon 1884, 116)

Webster also records fissile rock near the surface and correlates it to the Lower Purbeck.

> Immediately under the soil, which seldom exceeds a foot in depth, is a series of thin beds, all together about 3 feet thick, called slate by the quarry men, which split readily into layers from an inch to half an inch in thickness. They

Figure 29. The Middle Purbeck Beds are the best known source of roofing in Purbeck and Portland. A modern quarry near Langton Matravers which clearly shows the different beds of stone (© Terry Hughes).

Table 8. Lower Purbeck Bed names in Porryfield quarry, Portland.

Unit	Bed	Thickness in metres (feet)
	Slatt marl and Burr	1.0 (3.2)
	Clay and Marl	1.65 (5.4)
	Slatt and Marl	0.37 (1.2)
	Clay	0.7 (2.3)
	Thin Slatt	0.25 (0.8)
	Shingle and clay	0.4 (1.3)
Lower Purbeck	Thick Slatt	1.0 (3.2)
	Shingle with clay above	2.1 (6.8)
	Clay	0.3 (0.9)
	Aish Tier	2.8 (9.0)
	Black Dirt Bed	0.25 (0.8)
	Top Cap	2.0 (6.5)
	Black Dirt	0.1 (0.3)
	Skull Cap	0.5 (1.6)
Portland Stone	Roach	

> consist of limestone of a dull yellowish colour, extremely compact, and entirely without shells (at least I have not seen any in it). In its aspect it considerably resembles those compact varieties of Purbeck stone in which the remains of shells are not visible. These are what I consider to be analogous to the beds in the lower part of the Purbeck limestone, seen at Warbarrow and Mewp Bays above mentioned. (Webster 1826, 41)

Woodward recorded:

> Thin limestones in the Lower Purbeck Beds on Portland, known as the Slatt Beds and some of the overlying fissile limestones have been used for roofing. (Woodward, 1895, 318)

It is unclear whether the term slatt always implies a fissile rock suitable for roofing. Arkell and Tomkieff (1953) clearly link the meaning to slate but it is also a general term for the Lower Purbeck Beds above the Soft Cap (Ian West, personal communication) [8] and Arkell (1945, 167) has indicated that this meaning does not necessarily apply in the Purbeck beds. He states: 'Slate, Slatt, Slattern. The rock so called in Purbeck and Portland, however, is useless shaley limestone with a slaty fracture (fissile along the bedding-planes); as Damon describes it "shivered stone".' It is unlikely that all thin layers would be sufficiently strong for roofing, but some were. Table 8 indicates how much of the Lower Purbeck may have been fissile in one quarry, Porryfield (Cope and West 1969, A63–64).

North of Upwey, the Purbeck Beds have apparently not been quarried for roofing, but sources near Tisbury in the Vale of Wardour, north-east of Salisbury have been used in Wiltshire (see Cotswolds section, p58).

Market

Although there were many stone slate delves in the region limestone roofs are not common in West Dorset (Clifton Taylor 1962) and Somerset. Nonetheless, there are probably enough roofs to sustain at least one quarry. Because there is such a variety of stones which have been used there is a need to review how many of these are essential to conserve the local roofscape. All the Jurassic stones weather to approximately the same grey colour although, because some are more susceptible to spalling, especially at their edges, many roofs continue to have a yellow component. Colour aside, roofs do show considerable variation because of the texture of the individual slates and the roughness of the roof surface.

Buildings in south-east Dorset have been roofed with Purbeck slates from both sources, although outside the Isles it is not a major regional roof type. Even in Wareham, less than ten miles from the Purbeck quarries, there are very few Purbeck slates and weight and the difficulties of transport would have limited penetration further northwards. Although buildings in these areas may periodically need a lot of slates for their maintenance this demand will be intermittent. To the north of Purbeck, at Wareham for example, stone slates are commonly used at the eaves of clay tile and thatched roofs but these will only need a small supply for occasional re-roofing (Fig 30).

Within the Isle of Purbeck, stone roofs are a fundamental feature of the vernacular tradition and have been used on buildings of every type (Figs 28 and 31). All the

Figure 30. Beyond the Isle of Purbeck, as here at Weybridge, Surrey, stone slates are used at the eaves beneath tiles, thatch and metamorphic slates (© Terry Hughes).

quarry owners report a continuing demand which they are generally only able to supply with difficulty or not at all. The shortfall is often made up with sawn product but even this fails to prevent the loss of roofs to other products.

Production

Potentially the quarries listed in Annex A (p89) could supply stone slates if suitable rock occurs.

Lower Lias
Near Somerton, Station Quarry at Charlton Mackerel and Tout Quarry at Charlton Adam are working the Blue Lias but do not produce roofing.

Upper Lias
Slates are intermittently available from the Ham Hill area. The owners of Ham Hill quarries have shown interest in the possibility of producing sawn slates. These would need to be tooled over the surface and edges but, since this stone slate was often tooled to reduce its thickness and flatten it, this may be visually acceptable (Fig 32). Early trials have shown that a very good match for old slates can be produced, and durability tests are being carried out.

Corallian
There is no roofing produced from this formation.

Forest Marble
In Dorset, there is no production in the Forest Marble at present but it is delved in Wiltshire and the Cotswolds, from where it is sourced as available. One delph at Stalbridge, near Sturminster Newton, has recently been opened in the Forest Marble Formation and may be able to produce slates. These are likely to be quite rough-surfaced, as is traditional in the immediate area. There is a possibility that production may be re-established in the Sherborne area if a sufficiently large market can be confirmed. The existence of suitable rock has been proved on the Sherborne estate (David Jefferson, personal communication).

Purbeck Beds
There is little demand for stone roofing in Portland nowadays and consequently no production. If a source of suitable rock was available it could help alleviate the supply problems in Purbeck and further afield in Dorset. Recent output from the Slatt bed of Albion Quarry has been about 20–25 mm thick but of small size (Ian Grey, personal communication).

Historically, the coastal exposures were worked as open quarries on the Isle of Purbeck to obtain Portland stone, from Seacombe Bottom to Tilly Whim. On the higher ground of the Downs, there are many remains of small open workings or ridden pits. Tilly Whim Hill, above Swanage, was extensively worked and there are traces of old underground workings or quarr, some filled-in but others well preserved, especially north of the B3609 where the dip slope was worked from Swanage to Orchard, south of Church Knowle (Arkell *et al* 1947). Later, almost all production, including stone slates, was from underground workings. In 1877, there were at least 92 mines being worked (Velacott 1908).

The methods employed in the 1940s were described by Arkell.

> The method of mining the stone is by a steeply inclined shaft to a depth of a hundred feet or more, from which galleries are driven underground along the bedding, in the direction of strike. The blocks of stone are brought to the surface by a winch worked by a horse or pony … although the Purbeck building stone still enjoys a steady market, some of the thinner seams, formerly used for roofing, are now sold for crazy paving. (Arkell *et al* 1947, 106)

The roofing beds are at different levels in different locations. In old documents they are described as both so shallow that in the underground workings plant roots

Figure 31. Purbeck stone is capable of being split into very large slabs, although on normal roofs, sizes seldom exceed four feet wide (1.3 m). Large stone slates like this, sometimes called covers, are commonly seen close to quarries in many parts of England where they were usually used for small buildings such as pigsties. A Yorkshire example is shown in Figure 103 (© Terry Hughes).

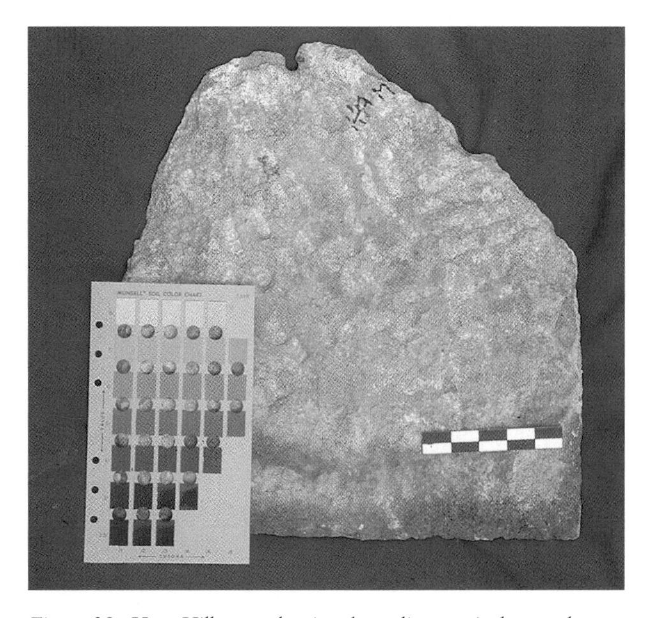

Figure 32. Ham Hill stone showing the tooling required to produce a reasonably flat surface. Munsell 7.5YR (© Terry Hughes).

Figure 33. Purbeck stone. In some quarries slabs are sawn off the face of masonry blocks to produce roofing with a natural surface. They are then edge-dressed. To look authentic the edge dressing must be done very subtly. It is all too easy to produce edges which are too straight, giving the roof an artificial appearance. Some quarries also saw flat slabs which are surface and edge-dressed (© Terry Hughes).

hung down from the roof, but also as deeper than the current open workings. Nowadays, quarrying operations are all above ground and the quarries are working in two areas: north of the B3609 and south-west of Acton. This is part of a plan to concentrate the activity so as to limit the visual impact.

All the quarries are worked with modern excavation machinery, which risks damaging thin fissile beds, and all have diamond saws for processing the block. Beyond that, the level of organisation and scale of operations vary from small family businesses involving two or three people to substantial companies with permanent office and sales staff.

Supplies of fissile rock are small and intermittent and, since the rock must be worked progressively through the beds, there is little that can be done to change this. There is simply not enough fissile rock available to provide an assured supply. Although this was probably always the case, in the past it was easier to obtain stone slates by shopping around, because different quarries would tend to be working different beds at any one time. Now, because the quarries are being concentrated in a smaller area, especially those near Acton, they are more likely to be working the same beds at the same time.

In early 1997, any rock suitable for roofing would have come from the Thornback or Wetson Beds or the Downs vein. However, two quarries have worked through the Cinder into the Button Bed which it is assumed would be suitable for stone slates.

When suitable rock is available, some quarries are prepared to provide slates to a cutting list and one will even set out and cut hip slates to size. While such initiatives will be popular with some roofers and may keep down the roofing cost, there is a risk of losing the natural variability of the roofs, which is dependent on the random nature of the unworked stone.

To overcome the supply difficulties, it has long been the practice to produce roofing from the top and bottom slices sawn from a non-fissile block in the preparation of masonry products (Fig 33). The resulting slates have a 'riven' top face, a sawn bottom face and the bottom edges and about half of the sides are then dressed with a hammer to produce a traditional appearance. If this is done well, it can give a realistic result provided the underside is not to be seen. One quarry saws roofing from block and tools the top face as well as the edges.

It is not known whether all the rocks used for sawn slates will be adequately durable, although all the quarries claim to use only the most durable stone for this purpose, Down's Vein, Wetson, Thornback, Roach and Grub. It would be wise to include in any specification for such stone slates a limitation on which beds are to be used, unless or until convincing evidence of durability is available. It is important to allow newly quarried slates to stand for some time before putting them on the roof, otherwise they can be damaged by frost. Traditionally quarrying is restricted to the warmer months to avoid frosts ruining the block.

Other sources

Reclaimed slates from redundant buildings or other roofs are the major source of supply for re-roofing. Stone roofing specialists and building owners will stockpile second-hand material whenever it is available. While this is a solution to the problem of conserving the most important areas and buildings, it cannot be sustained. Purbeck roofs taken off some houses on a gunnery range, to avoid them being damaged, sustained the market for some time but such occasions are rare and unrepeatable. Meanwhile every roof that is changed to some other product degrades the region. The idea of Corfe Castle, Langton Matravers and Worth Matravers without their stone roofs is sad to contemplate but this may ultimately be the fate of the area unless sufficient new material can be fed into the repair cycle.

In principle, Purbeck roofs could be repaired with similar stone slates from outside the region, either new or reclaimed but there is no evidence that this is happening. The Cotswolds delves are the obvious choice but as these slates are smaller on average, and often thinner, their use could result in a radical change of appearance.

It has been suggested that stone slates of similar geology and appearance are available from France. This has not been confirmed.

Summary

Stone roofs are an important part of the regional roofscape which is only being conserved with difficulty and often at the cost of unprotected roofs. This region has a diverse selection of roofing stones but no local production except on the Isle of Purbeck, which faces an intractable problem. The quarrying companies feel that there is simply not enough suitable rock to supply the demand for new material to be fed into the repair cycle. This problem is being aggravated by the policy of restricting the location of quarries to satisfy non-quarrying objectives. Even the use of sawn slates fails to prevent the loss of roofs to other products. The policy for this area may have to be based on very careful management of a very restricted supply. It may even be necessary to accept recycling from less important roofs onto more important ones, such as listed buildings and those in conservation areas. The only other potential alternatives are to accept somewhat dissimilar stone slates from a Cotswold source or to import slates from France if they are suitable. As the Cotswolds may not provide a long-term solution because this region has its own supply difficulties, the latter option needs to be investigated.

THE COTSWOLDS

> It may be interesting to notice that the occurrence of these so called 'slate beds' of the Oolites ... is in almost every instance a phenomenon of very local character. The presence of such beds depends on the existence, in a rock mass of a finely laminated structure, of a due admixture of calcareous and arenaceous materials; and as the necessary conditions for their formation can scarcely be expected to prevail over any extended district, we are not surprised to find that the peculiar features of such rocks are only found over comparatively small areas; the 'slate' passing within very short distances either into loose sand on the one hand, or into solid limestone rock on the other. This is the case alike with the Stonesfield, the Collyweston, and other similar 'slates'. (Judd 1875, 5)

For convenience, this section covers those parts of Wiltshire, Gloucestershire and Oxfordshire in which stone slates of Jurassic age are used. The other stone slates of Gloucester are included in the Welsh Marches and the Bristol and Forest of Dean sections.

Geology

> Stone tiles are principally raised on the Cotswolds in different parts; the best are prepared at Miserdine, Bisley, Beaverstone, Charlton in the parish of Tetbury, Hampton-field, and Ablington in the parish of Bibury. The colour of these are yellow, or grey, but another sort of red grit [9] is dug at Iron Acton, and some adjoining places; as these, however, separate into thick lamina, and of course require strong timbers for their support, they are less eligible than the former; all stone tiles indeed, on account of their weight, are not so much used as those which are burnt from clay, or the light blue-slates from Wales, where not prevented by the expence of carriage. (Rudge 1813, 24) [10]

The stone slates of the region exhibit a variety of lithologies including fissile, fine-grained oolites, sandy limestones, calcareous fine-grained quartzose sandstones and siltstones (Fig 34, Table 9). This is the consequence of the variety of sedimentary environments which existed in the region in the roughly 50 million years during which the sediments were laid down, and the variety of sources from which they were derived. Unfortunately, because the various lithologies (the rock types and their characteristics) change over quite small distances and are repeated vertically, it has not proved easy for geologists to define the geological succession and to correlate the various rock types across the regions. This problem has resulted in the creation of many local geological names which have been adopted for a period and then abandoned following new research. Consequently, the names Stonesfield Slate and Cotswold Slate have been mistakenly applied to various outcrops and the delves where they were worked, so care is needed when using older geological reports and books.

Within the region, two major types of stone slate are identified on the basis of their fissility and methods of manufacture: presents and pendle. Presents can often be found 'pre-split': normally they are delved from near-surface deposits and may not need to be physically split. They have been widely produced throughout the region. Pendle is obtained from deeper deposits which have not been exposed to periglacial influences and postglacial weathering, and is split by exposing it to frost action. It is important that the block (called log, lens or pot lid) is not allowed to dry out before frosting since this results in changes within the stone which are not only irreversible, but also prevent the stone splitting when frosted. Slates formed in this way have been known generically as Stonesfield slates after their most important source where they have given their name to the formation in which they occur, the Stonesfield Slate. Strictly though, this name should be reserved for slates from the Stonesfield area.

Whether a particular fissile rock is pendle or presents depends on the nature of the 'cement' which binds the rock grains, and Boneham and Wyatt concluded that the beds of Stonesfield Slates are better regarded as a recurrent and sporadic facies [11] rather than a formal lithostratigraphic unit. Because of this, a particular source of potentially pendle rock may not in practice be frost-splittable, or may contain both types. This is especially the case in other areas where pendle slates were produced, such as in the Taynton Stone north of Naunton and its southern extension, the Througham Tilestones, at Bisley and Througham. The delves around Naunton mainly work the Eyford Member of the Fullers Earth but this is overlain by Taynton Stone and from time to time this has been worked to produce both pendle and presents.

Pendle slates tend to be thinner, flatter and smoother than presents, although in the heyday of the Stonesfield industry two or sometimes three grades or thicknesses were produced. The thickest, which are no thinner than presents, can still be seen on the roofs of sheds in Stonesfield village. (Care needs to be exercised in interpreting roofs in the Stonesfield area as many roofs have been removed and replaced by other Cotswold stone slates.) Presents weigh about 160 kg/m^2 (1.5 tonnes per square [12]). Because the fissility of both types varies and because they have all, in recent times, been used over a wider area than their geological occurrence, the local roofscapes are more diverse than might be expected.

Table 9. Sources and historical stone-slate production centres of the Cotswolds region.

	Age	Rock Unit	Locations and local names	
	Cretaceous/Upper Jurassic	Durlston Formation Lulworth Formation	Swindon (Purbeck slates)	
Upper Jurassic	Kimmeridge Clay Formation Ampthill Clay Formation			
Upper Jurassic	Corallian		Buckland Warren (Pusey slates)	
Upper Jurassic	Upper Oxford Clay			
Middle Jurassic	Middle and Lower Oxford Clay Kellaways Formation			
Middle Jurassic	Great Oolite	Cornbrash		
Middle Jurassic		Forest Marble	Poulton, Ampney Down, Burford to Lechlade, Bradford on Avon, Neston Park, Tetbury, Chevenage, Crudwell	
Middle Jurassic		White Limestone including Hampen Marly Beds in the centre and north	Cheltenham to Cirencester	
Middle Jurassic		Taynton Stone in the north	Burford to Enstone (Fulwell slates), Soundborough.	
Middle Jurassic		Througham Tilestone Formation in the south	Througham (Bisley slates)	
Middle Jurassic		Sharps Hill Beds in the east	Shipton under Wychwood	
Middle Jurassic		Stonesfield Slate in the north	Rag – false bedded oolite, Top Soft or Marly Beds, **Overhead, Upper Head or Potlid** **Race or Manure** **Lower Head** Softstuff, Bottom Stuff or Block	Stonesfield
Middle Jurassic		Fuller's Earth including the Eyford Member	Andoversford to Lower Slaughter, Windrush Valley, Soundborough, Brockhill, Sevenhampton Common, Chalk Hill, Stow on the Wold	
Middle Jurassic		Chipping Norton Limestone in the north	Moreton in Marsh	
Lower Jurassic	Inferior Oolite		Snowhill, Kineton	
Lower Jurassic	Bridport Sand, upper part Upper Lias			
Lower Jurassic	Middle Lias including the Marlstone Rock		Chacombe	
	Lower Lias			

Price found that presents dominated south-west of a line from Gloucester to Swindon, and pendle was most common in west Oxfordshire, but there was a large area of mixed use between the two, confirming that 'the production of pendle was not restricted to Stonesfield' (Price 1995, 51).

Middle Lias including the Marlstone Rock

Woodward recorded the production of stone roofing at Chacombe near Banbury.[13]

> Thin flaggy beds in the Marlstone at Chacombe near Banbury have been employed as tile-stones. A dark stone from Byfield (probably Marlstone), was formerly worked and cut into squares for paving halls. (Woodward 1893, 295)

Inferior Oolite

At Snowshill quarry (now Hornsleasow), and Kineton this formation has produced stone slates.

> Slaty beds were formerly worked for roofing purposes on the summit of the Cotswold Hills, south-east of Snowshill at a spot known as Hyatt's pits; and there also were 'slate quarries' further to the south-east. The stone-tiles have been used in the village of Snowshill together with Kyneton (Keynton) Slates from the Great Oolite. The former are thick and heavy. Somewhat similar beds are exposed in the freestone quarries near Longborough, and there can be little doubt that these beds belong to the upper part of the Lower Freestone, or to the horizon of the Oolite Marl. (Woodward 1894, 140)

Fullers Earth Formation

The Eyford Member of the Fullers Earth Formation outcrops along the high ground from Andoversford to Lower Slaughter (Fig 35) and in the Windrush valley, producing roofing at Brockhill, Goldhill Quarry Naunton (Fig 36), Soundborough, Salperton Downs, Sevenhampton Common, Chalk Hill and Stow-on-the-Wold.

Figure 34. A Forest Marble roof at Bibury, Gloucester (© Terry Hughes).

Figure 35. Fullers Earth Eyeford Member stone slate at Lower Slaughter (© Terry Hughes).

Figure 36. Fullers Earth, Eyeford Member stone slate. When stone slates are sawn to size and then dressed there is a tendency for the long edges to be too straight, giving the roof an excessively regular appearance. A more traditional appearance can be achieved if the edges are 'broken' in a more irregular way (© Terry Hughes).

Sharp's Hill Beds

Along the escarpment to the south-east of Shipton under Wychwood a narrow outcrop of fissile sandy limestones were worked in the past for roofing. They have been included in the Sharp's Hill Beds although this is questionable and they have also been incorrectly described as Stonesfield slates. The outcrop is narrow and the delving was on a small scale for local use.

Stonesfield Slate

> But before we leave of materials for Building, we must not forget that the Houses are covered, for the most part in *Oxford-shire* (not with tiles) but *slat-stone*, whereof the lightest, and that which imbibes the water least, is accounted the best. And such is that which they have at *Stunsfield*, where it is dug first in thick cakes, about *Michaelmass* time, or before, to lye all the winter and receive the frosts, which make it cleave in the spring following into thinner *plates*, which otherwise it would not do so kindly. But at *Bradwell* (near the Grove) they dig a sort of *slat-stone*, naturally such, without the help of *winter*, and so strangely great, that sometimes they have them of seven foot long, and five foot over: with these they commonly make mounds for their Closes, and I have seen a small *hovel*, that for the whole covering has required no more than one of these *stones*: and some of them are so hard and close a *texture*, that I have known them by *Painters* of very good skill, preferr'd before *Marble* for grinding their *colours*. (Plot 1677, 77)

The Stonesfield Slate lies above the Fullers Earth Formation in the Taynton Stone. Strictly speaking the name Stonesfield Slate should be restricted to the area around Stonesfield village. But in the past, the name has been applied to geologically similarly rocks including whether or not it was exposed at the surface and, consequently, presents. The pendle mined at Stonesfield and frosted for roofing was the most desirable of the Cotswolds stone slates because it split so thinly (Fig 37) and therefore needed lighter roof timbers.

> No place is richer in geological associations than Stonesfield. Its very name savours of the earth's crust, and indeed the ground has yielded, since the Roman occupation at any rate, the shelly and oolitic tile-stones known as Stonesfield Slates, while the fossils have attracted attention for 200 years or more. Plot speaks of the 'Flat-stone' of Stunsfield, and the beds have been worked along the sides of the valleys to the south, south-west, and west of the village (south of Hillburn Farm). The open works, with levels driven into the hill-sides, having for the most part exhausted the accessible material, shafts have been sunk in and about the village to various depths, ranging from 20 to nearly 70 feet, according to the thickness of the overlying strata. The village itself and the lands to the south and west, are therefore riddled with shafts and levels, and great heaps of waste material remain to attest the work that has been done.

Figure 37. Stonesfield Slates at Stonesfield. By exposing the stone to frost action the Stonesfield Slate could be split thinner than other Cotswold stones (© Terry Hughes).

Figure 38. Tayton stone. Slates were quarried from the Taynton Stone near Enstone, at Fulwell and Cleveley, and near Burford producing presents. These were known generally as Fulwell slates. West of Stow-on-the-Wold the Taynton Stone was occasionally worked above the Fullers Earth Formation producing presents and frost-split pendle (© Terry Hughes).

Figure 39. Througham Tilestone. These pendle slates, known as Bisley Flats, were quarried between Cirencester and Stroud (© Chris Harris).

The 'slates' are of three kinds; brown calcareous sandstone, grey and slightly oolitic calcareous sandstone, and blue and grey oolitic limestone. The strata yielding these materials are of variable nature, usually from 2 to 3 feet in thickness and not exceeding 6 feet; and from one to three layers, in different places, yield suitable stone.

The only partially open working that I have seen, was situated at Red Hill, on the eastern side of the valley between Fawler and Hillburn Farm. Here the 'slate' was exposed, although worked underground in the hill-side. The following section was shown:

			Ft.	*In.*
Great Oolite Lower Division	Alternations of oolitic limestone and marl		3	0
	Marl with thin films of sandy limestone: Pecten vagans and Rhynchonella concinna		6	0
	Roof Bed: grey oolitic and sandy limestone (that does not split, and is of no use)		2	0
	Stonesfield Beds	Shaley bed	0	6
		Stone worked for 'slate'-grey sandy and oolitic limestone	1	0
		Shale bed		

The slate mine (belonging to Mr Barrett) on the eastern side of the village of Stonesfield is one of the deeper pits. The shaft is 66 feet and the blocks of stone are raised with a windlass and a stout rope. (Woodward 1894, 310)

Sorby stated that 'the fissility of the Stonesfield Slate is, in great measure, due to minute laminae derived from *Ostrea* and Brachiopoda' (Sorby 1879, 85).

Chipping Norton Limestone

This rock, which is exposed only in the north of the region, was worked for roofing at Moreton in Marsh, Eyford and, in a small way, at Stonesfield 'two thirds of a mile south-east of Stonesfield Church' (Arkell 1947, 139).

Taynton and Througham Formations

Taynton Stone was occasionally fissile, producing presents from near Enstone, at Fulwell and Cleveley, and near Burford. They were known generally as Fulwell slates (Fig 38). West of Stow-on-the-Wold the Taynton Stone was occasionally worked above the Fullers Earth Formation producing presents and pendle.

Bisley and Througham

The Througham Tilestone beds were once widely exploited for the Bisley slates (also known as Bisley Flats) which are pendle (Fig 39). They extend from Minchinhampton to Througham and pass into the

Taynton Stone Formation in the direction of Burford. In the quotation below they are incorrectly described as Stonesfield probably because they could be split by frosting. Later studies have separated the two formations.

> Stone slates or tiles have also been obtained from Miserden (Miserdine), Rendcomb and Nettlecomb near Birdlip; and Prof. Hull states that they have been worked on the north-east side of Oakridge Common and at Battlescomb east of Bisley. (Woodward 1894, 140)

> At Througham field, north-east of Bisley, and south of Troughham or Druffam, we find quarries opened for the working of Stonesfield or Bisley slates. The Stonesfield Slate series is overlaid by 10 ft of current bedded oolite, which is quarried for building stone, for wall stone and road metal. The details of the underlying beds vary very much, but the following section noted in Mr Freeman's quarry (in company with Mr J H Taunton) affords a good example of the series: there the oolitic limestone is not seen:

		Ft.	*In.*
Stonesfield Slate	Soil		
	Oolitic, sandy stone, yielding the best 'slate' but now for the most part worked away	2	6
	Calcareous sandstone, used for wall-stone and road metal		
	Soft calcareous sandstone with scattered oolitic grain	4	6
	Hard calcareous sandstone: building stone		
	Fissile sandy beds	2	6
	Hard irregular earthy sandstone, obscurely oolitic; with fossils	1–2	2–6
	Fissile micaceous sandstone with Trigonia impressa. Slate Bed. The top 4 in. will never split; the next 9 in. forms good slate; the lower part is not so good	2	0
	Fissile sandy beds	1	6
Fullonian	Clay		

> The 'slate' does not exhibit planes of division in the quarry. It is never more than 18 inches thick, and this occurs at slightly different horizons. (Woodward 1894, 281)

Hampen Marly Beds

Some of these fine-grained oolitic limestones have been delved for roofing between Cirencester and Cheltenham. In some earlier texts the limestones in this area have been attributed to the Taynton Stone but they are now regarded as part of the Hampen Formation.

Forest Marble

The Forest Marble has been the most extensively delved source of presents, producing a coarser slate than other beds in the region. The main production areas were between Burford and Fairford, at Aldsworth, Holwell and Shilton; around Poulton and Ampney Down; around Tetbury (Fig 40) at Avening, Beverstone, Chevenage and Charlton and east of Bath between Bradford-on-Avon, Monkton Farleigh and Atworth. They continue to be

Figure 40. Forest Marble from the Tetbury Stone Company's delph (© Terry Hughes).

produced at Crudwell east of Tetbury and are expected to resume at another location at Tetbury in the near future.

CIRENCESTER TO FAIRFORD AND BURFORD

The false-bedded layers of Forest Marble were well shown in the delph at Crickley Barrow north-east of Coln St Denis, where the beds were seen to rest on the Great Oolite. The section was described by Hull.

> At Talland Quarry I noted the following section B:

		Ft.	*In.*
Forest Marble	Flaggy and rubbly stone	4	0
	Clays and thin fissile beds or oolite and gritty limestone	8	0

> On the thin gritty layers, many trails occur; and among the fossils I obtained *Pecten annulatus*, *P. lens*, *P. vayans*, *Ostrea sowerbyi*, and *Rhynchonella*. (Woodward 1894, 369)

Also between Cirencester and Fairford the so-called Poulton slates were produced.

> At Poulton Quarry we find beneath from 2 to 4 feet of brown clay, an alternating series, 12 feet thick, of obliquely-bedded bands of oolitic limestone and grey clays. The limestone or 'blue stone', occurs in thin flags which are largely employed for roofing-purposes, under the name of 'Poulton Slates'. (Woodward 1894, 370)

> Some of the fissile and obliquely fissile masses of shelly and oolitic limestone are quarried for 'slates' and planking [14] south of Burford Signett and Holwell. The details vary in each opening. The stone beds alternate with racy [15] clays and are exposed for a depth of from 6 to 12 feet. (Woodward 1894, 369–70)

TETBURY

> At Chevenage, near Tetbury, roofing-tiles have been obtained from the Forest Marble. The beds here, as at Poulton, occur in thin layers, the surfaces of which abound with specimens of *Ostrea*, *Pecten*, *Lima*, *Rhynchonella*, &c. all much compressed and distorted. (Woodward 1894, 365)

In recent times the main sources of Forest Marble stone slates have been to the west of Tetbury and at Chevenage but formerly delves were also worked at Avening, Beverstone and Charlton. Currently delves about one mile west of Crudwell and on a small scale at Down Ampney are producing Forest Marble slates.

BRADFORD ON AVON

> Above the Bradford Clay near Bradford-on-Avon, we find beds of shelly and earthy limestone, with much clay and marl, and thin leaves of sandy limestone with curious tubicolar markings. Locally I saw no evidence of any thick series of limestones, and no evidence of the Hinton sand and sandstone. The beds at Westwood, and again at Frankley north of Bradford-on-Avon, are mainly argillaceous, with occasional bands of stone: so that the shallow wells sunk into this formation obtain but limited and uncertain supplies of water.
>
> Further north, between Monkton Farleigh and Atford[16], the stone-beds are again of importance. Lonsdale noted 10 feet of 'shelly limestone, split into thin layers obliquely to the plane of stratification', at the Wormwood quarry [17], on the high road from Bath to Devizes. Here there was no evidence of the Bradford Clay, for these beds rested directly on the Great Oolite freestone. North of Atford, an old quarry showed the following strata:

		Ft.	*In.*
Forest Marble	Brown sands with fissile concretionary of grey calcareous sandstone.	6	0
	Thin flaggy and sandy beds, and false-bedded shelly oolitic limestones	13	0

> Water being held up at the base of the pit, indicated a clay foundation. In this neighbourhood some of the beds have, in old times, been used as stone-tiles. (Woodward 1894, 355)

No delves are producing stone slates in this area at present. A delph opened at Atworth during 1999 proved to have insufficient fissile rock to be commercially viable and has been closed.

Corallian

Pusey slates were produced from the Corallian at Buckland Warren and Pusey on either side of the A420 (Fig 41). There are no stone slate delves in the area today.

Lower Purbeck Beds, Lulworth Formation

Around Tisbury in the Vale of Wardour, roofing was produced on the north-east side of the town: '[on] Ladydown, delves have long been worked for the sake of tilestones ... a fissile stone, yielding *cyrena* and remains of fishes' (Woodward 1895, 272).

There are no delves producing roofing in the Vale of Wardour at present.

> At Swindon the Lower Purbeck Beds have yielded tilestones used in some of the old buildings in North Wilts and at Devizes. (Woodward 1895, 318) [18]

The Purbeck and Portland Formations are present as an isolated outcrop at Swindon and were worked at the Town Gardens quarry, mainly for the lower Portland stone which is unlikely to have been a roofing source. By 1940, the quarry had been surrounded by the town but the geological succession was reviewed in that year (Sylvester-Bradley 1940, 349–72). From this it is concluded that the stone slates probably came from the Swindon Flags at the surface. The quarry is now closed but is preserved as a Site of Special Scientific Interest.

Market

In an extensive survey carried out by Price in 1995, it was found that, within the region of historic use, stone roofing still comprised 28% of all roofs. Within historic town and village centres the percentage was, of course, much higher. This is a substantial market, certainly enough to sustain an industry even without the significant numbers of new buildings which use stone slates. Within this market Price found that pendle slates predominated in the east, presents in the west and there was a large area of mixed use between them.

Of nearly 1000 buildings surveyed which were thought to have been originally covered with stone slates, 42% had lost all, or a substantial part, of their slates. This was partly due to a lack of planning control: 62% of stone roofs were on unlisted buildings. This is inevitable if the use of reclaimed slates is permitted.

Price also examined the effectiveness of conservation controls and the support provided through grants and found them to be inadequate.

Figure 41. Pusey slates were produced at Buckland Warren and Pusey from rock of Corallian age (© Chris Harris).

Some local authorities have grants for re-roofing with stone slates. All agree that the available funds are inadequate to stem the loss of roofs or use of reclaimed slates. (Not all would wish to do the latter.)

An Environmentally Sensitive Area scheme exists for the Cotswolds Hills. Although this is primarily directed at the conservation of traditional farming systems which have contributed to local, distinctive landscapes, habitats and historic features, it also includes a capital works grant scheme for the conservation of drystone walls and the restoration of farm buildings using traditional materials. The managers of the scheme have reported that funding is often insufficient to conserve stone roofs, especially for farm buildings which have no use in modern farming systems. The result is that grants are often applied to re-roofing with imitation stone slates.

Production

Except at Stonesfield, production has always been in open delves, and even here the earliest workings were near the surface. It was only when these were exhausted that mining developed, motivated by the economic advantage of such thin-splitting rock and the lighter roof structure it required. Woodward (1894, 482–86) described their manufacture.[19]

> At Stonesfield the slate is now mostly obtained by means of shafts, there being one mine worked on the level. The shafts are from 20 to 70 feet deep.
>
> The beds worked are from 2 ft. 6 in. to 3 feet thick, and they yield brown sandy slates, hard grey and slightly oolitic slates, and blue and grey or brown oolitic slates. In size they are obtained 1 ft. 6 in. square and less. The blocks of fissile oolite, fine-grained calcareous sandstone with courses of oolite, and calcareous sandstone, when dug out, must be kept damp, and then exposed to a winter's frost. After that they can be split up evenly at any subsequent time. If they have been allowed to dry before the frost comes, so that the 'sap' is dried up, no frost will move the layers, and the rock is then 'bound'. Formerly it was the custom to begin digging the stone about Michaelmas time or before; now the men commence about the end of October, and work till the end of January. When dug and raised, the stone is banked up and covered with earth, to keep the moisture in, 'until a frost comes.[20] In dry seasons, the quarrymen even water the blocks to keep them moist. It requires three or four good frosts to act upon the stone; hence a mild winter is disastrous to these toilers. Sometimes the stone is put out and then covered up, again and again, until a sufficiently long frost is experienced. As a rule a week's frost is needed.
>
> The stone can be split up any time after it has been 'frosted', and of course it can be dressed any time after it has been split. Splitting is usually commenced about the middle of February.
>
> The only qualities in the slates are the first and second, the former being thinner and the latter thicker, though both may be composed of the same kind of stone. The more oolitic slates as a rule are a trifle thicker than the sandy slates, but no particular difference in quality and durability is recognized.

Sufficient slate was produced to occupy the slate makers, or slatters, throughout the year. Once frosted the splitting was completed with a hammer and any loose flakes were knocked off. Woodward (ibid) also recorded the technique further south.

> At Througham Field, where Stonesfield Slate[21] is worked, the blocks are improved by lying out all the winter – the second or third frost breaks them up. Slabs 6 feet square or even more are obtained. They furnish material for cow-sheds, mangers, bordering for gardens, and pitching for stables. There is not more than 18 inches of good tile-stone at this locality, but it occurs at slightly different horizons. I obtained some examples of the tools used at Througham Field, and these are placed in the Museum of Practical Geology.

Delving of presents is straightforward, only requiring the productive beds to be uncovered and the slates prised out (see Figs 4-6). If necessary, they were further split before they dried out, and then trimmed to size and shape.

The process of conversion to the finished slate was basically the same for pendle and presents. The thinner end was selected to be the top edge and the shoulders removed to reduce weight and to make them easier to lay. Generally, the edges are 'double' bevelled (Fig 42) which helps them to sit well on the roof, but this is not universal. In Lower Slaughter, Througham and Pusey, for example, some of the old roofs have slates with edges which are almost square. The peg hole was formed with a pick-hammer.

A review of the development and decline of the Stonesfield industry is included in Aston 1974, and of the Cotswolds generally in Price 1995. In *The Natural History of Oxford and the Valley of the Thames* (1677), Plot described an industry at Stonesfield based on the frosting

Figure 42. The Througham stone was widely exploited for roofing around Bisley and Througham. This example shows the double beveled edges typical of many Cotswold stone slates (© Terry Hughes).

Figure 43. Fullers Earth slate from Soundborough delph. Munsell 7.5YR (© Terry Hughes).

process well-established by 1676. The presents industry was at least as active at that time. Both were benefiting from the economic expansion which followed the ending of the Civil War (Strange, in Hughes 1996, 20) and continued to respond to market influences, including the agricultural depression in the middle of the eighteenth century. Between 1800 and 1860, the population of Stonesfield had risen by nearly 70% to 650 persons, about 130 of whom were involved in slate production. This was the peak of the expansion. During the next 50 years the Stonesfield industry contracted and by 1911 production had ceased. A similar decline is recorded throughout the Cotswolds, with periods of fitful production incidental to other quarrying, or none at all during the middle of the twentieth century.

Since the 1950s, a series of revivals has taken place in various delves producing presents but these have struggled to sustain themselves. During recent years, a few delves have produced slates, but the output has been insufficient to supply the market without problems. The scale of the industry was so small that when any one delph ran out of suitable stone the disruption to the market was severe: this created a reputation for unreliability and provided a reason to justify substitution by other products.

In 2000 seven companies had the potential to produce slates at least from time to time. Four delves recently started production: at Tinker's Barn near Naunton (Cotswold Stone Quarries), in Goldhill Quarry near Crudwell (Cotswold Stone Tile Company) [22], Grange Hill Quarry (Natural Stone Markets) and at Soundborough quarry (Soundborough Quarry Company, Fig 43). Intermittent, small-scale production continues at Down Ampney.

Figure 44. Sawn-edge slates can have a very unnatural appearance unless the edges are subsequently dressed to a traditional finish (© Terry Hughes).

Further developments are in prospect at Tetbury Stone Co Ltd. All of these are, or will be, working presents. Annex A lists the quarries' contact and production details (p89).

Currently there are no proposals to work pendle because natural frosts are unreliable, but an artificial frosting process has been developed by the Cotswold Stone Tile Company.

Today, all operations are open delves. Overburden is removed and rock quarried with hydraulic excavators. Selection of suitable rock for slates is still a hand process but, in an attempt to improve productivity and to hold down the product price, some manufacturers are using saws to cut slates to size. Also in some delves they have tried to improve supplies by using thicker-splitting rock. Neither of these initiatives have been without problems. Sawn slates look artificial unless they are carefully dressed to remove the straight edges. When the edges of overly thick slates are dressed in a non-traditional way they produce large gaps on the roof. When both of these faults are present in the same slates, the resulting roofs look like badly-made concrete imitations and may be technically defective (Fig 44).

Traditionally, random slates were supplied by weight, based on a nominal (estimated) coverage per ton and the mix of sizes currently available. Sufficient slates were supplied to cover the roof area. This mix, when applied to the various roof shapes, resulted in a variety of coursing: some roofs would have more large margins than others. This is an essential feature of the vernacular tradition at both regional and local level. Today, some manufacturers try to help roofers by offering a bespoke service. At its simplest, enough slates are supplied for an existing battening pattern. This is very convenient during re-roofing. Under a more elaborate service, a battening plan is calculated for each roof slope and a mix of slates made up. The latter has the potential to reduce the regional and local variability of roofs, especially because slates are now supplied to the whole region from a small number of delves. To avoid this problem the coursing patterns could be randomised, a simple mathematical

Figure 45. French stone slates at Woodchester Mansion near Stroud. They were chosen as the best match for the original Forest Marble slates which were unavailable at that time (© Chris Harris).

procedure, or the approximate coursing pattern for the building's location could be included as part of the specification.

Other sources

IMPORTS

Slates have been imported from France by several companies in recent years (Figs 45 and 46). They are reported to have originated from several geological formations and their durability has been questioned by some specifiers without first testing them. One, the equivalent of a Forest Marble stone formerly supplied by The Completely Stoned Company, has been compared with a Fullers Earth Formation (Eyford Member) stone and found to be at least as good in respect of mechanical and durability factors.

The market has a preference for local products and a resistance to the use of imports on grounds of appearance. Objections are sometimes raised to the colour, thickness or the sawn edges. It should be understood that the French stone slate was not intended to be a substitute for the Fullers Earth. It was being sold as a replacement for the, at that time, unavailable local Forest Marble. For this reason it was selected for use on several buildings, including Woodchester Manor, a Grade I listed building, because it was a better match for the existing slates than others available locally at that time. While the colour and thickness of the French slates are inappropriate in some localities (just as some other Cotswold slates would be) the sawn edges were dressed off in a traditional manner to suppress the artificial look of sawn edges. Ultimately, the colour of a French roof will be the result of weathering

Figure 46. French limestone as used at Woodchester Manor. Munsell 7.5YR, centimetre scale (© Terry Hughes).

and vegetation growth and on French buildings its weathered colour is grey, the same as Forest Marble.

RECLAIMED SLATES

There is an active trade in second-hand slates and, regrettably, their use is supported by some local authorities. Price concluded that barns were the major source in 1995.

The reasons cited for use of reclaimed slates are cost, the unsuitability of modern slates both new and imported (edges, colour, thickness etc), the mistaken belief that old slates will be more durable than new and supply problems.

There is no evidence of reclaimed slates being brought in from outside the region.

Summary

The Cotswolds region has a substantial market for a variety of stone slate types, only some of which are currently available. Most importantly, there is no source of pendle slates. Manufacturers are trying to improve the supply of present slates by a variety of methods, including imports from France and some technical innovations. These are not all in the best interests of conservation.

Reclaimed slates are a major source of material for new roofs and renovations. This is driven by cost constraints (and profits for the vendor) and supply difficulties. In spite of grants and the high profile that stone roofing enjoys with conservation and amenity bodies in the region, stone roofs continue to be lost at a significant rate.

THE EAST MIDLANDS

> In the 'White Pendle', we have a prominent example of the limestone and slaty beds which I have found to occur at different points over a considerable area in the same position in the general section of the district. The calcareous nature of these beds and the slaty character of the so-called 'Colleyweston Slate', and indeed of the Colleyweston

Figure 47. A Collyweston laced valley. In these valleys the courses are swept up to butt against a diamond-shaped slate. Figures 127 and 128 show this process (© Terry Hughes).

Figure 48. Duston Slate from the Northampton Sand Formation (left) and Collyweston Slate from the Lincolnshire Limestone Formation (right), with centimetre scale (© D S Sutherland).

> Slate itself, I consider to be attributable to accidental and local causes. (Sharp 1879, 371)

Dr D S Sutherland has written a detailed review of the geology of fissile rocks in the region and the history of their use, and this section is mainly based on her unpublished report. [23] A number of stone beds have been worked for stone slates in the region but only one, at Collyweston, has been of a significant size (Fig 47). Production of Collyweston Slates has dwindled almost to extinction in recent years, and, in spite of the region's very favourable situation with unlimited potential supplies of stone, local people with the manufacturing skills and the existence of the Collyweston Stone Slaters Trust, specifically set up to promote stone roofing 17 years ago, there has been no progress in establishing commercial production beyond a small level of self-supply by roofing companies. This situation is expected to change soon.

Geology

> The *Slate* of this County is found either in thicker *Strata*, which being sprinkled with Water and exposed to Frosts, do readily cleave into such thin and eaven Plates as are fit for covering the Roofs of Houses: Or in thinner Strata which as they come out of the Earth are immediately fit for that use, without the Preparation above-mentioned. ... For the other sort, *Colly-Weston* is of great and ancient Fame. (Morton 1712, 109)

Only the rocks of the Jurassic Age have been exploited for stone slates in the region although, it is worth noting in passing, that the Swithland slate (which is metamorphic and outside the scope of this report) is another important local roofing type and is suffering from the same problems as the stone slates.

Historical records show that an industry was established in the Stamford region as early as 1286, but it is probable that several other fissile rocks were also being used locally at that time. Four which have been exploited for roofing are the Blisworth and Lincolnshire Limestones, the Rutland Formation (formerly the Upper Estuarine Limestone) and the Northampton Sand Formation (Fig 48). Occasionally more than one rock might be worked at a delph.

Other workers have given sources for stone slates in the region, sometimes attributing them to particular geological formations (usually the Collyweston Slate). Because of the complex structure of the geological succession in the region and the lack of grid references for the locations, it has not been possible to confirm all these geological attributions (Table 10).

Blisworth Limestone

Sharp described a quarry west of Oundle where pendle is 'split into thin flags or slates' (Sharp 1870, 5). In this context pendle is used in the quarryman's general meaning, fissile, and does not imply a frost-split slate, as it does in the Cotswolds region. A record of slatstone from Jerdle, used in rebuilding the Middle Gatehouse at Higham Ferrers, has been interpreted as a reference to Yardley Chase (Kerr 1925).

Rutland Formation (Upper Estuarine Limestone)

This formation exists at a number of horizons as thin limestone beds. It outcrops thinly around Kettering, Wellingborough, Northampton and Irchester. Southwestwards it thickens and was worked for building stone at Helmdon. Throughout its exposure it is laminated and suitable for roofing.

Archaeological sites in Northampton (Williams 1979, Williams *et al* 1985) have produced stone slates dating back to as early as 1250. They included examples from the Rutland Formation which is interpreted as coming from the Duston area, where the delves worked this horizon, as well as the Northampton Sand Formation.

Other sources of stone slates which are judged to have been in this formation include Pytchley (Morton 1712, 109), near Kettering, and Helmdon (Richardson 1925), near Brackley.

Lincolnshire Limestone Formation

> These divisions of Lower Estuarine Series and Lincolnshire Limestone, frequently shade one into the other by insensible gradations; and occasionally, at their junction, beds of fissile limestone occur, which constitute the

Table 10. Sources and historical stone slate production centres in the East Midlands.

Age	Rock Unit	Stone slate locations
Upper Jurassic	Oxford Clay	
	Kellaways Beds	
	Upper Cornbrash	
Middle Jurassic	Lower Cornbrash	
	Blisworth Clay	
	Blisworth Limestone (formerly the Great Oolite Limestone)	Oundle, Blisworth, Yardley Chase
	Rutland Formation (formerly the Upper Estuarine Limestone)	Pytchley, Hopping Hill near Duston, Helmdon? near Banbury
	Lincolnshire, Limestone Formation, includes the Collyweston Slate	Collyweston, Easton on the Hill, Kirby Lodge, Duddington
	Grantham Formation (formerly the Lower Estuarine Series)	
	Northampton Sand Formation	Duston, Harlestone Heath, Pitsford, Weston-Favell
Lower Jurassic	Upper Lias	
	Middle Lias including the Marlstone Rock	Chacombe near Banbury
	Lower Lias	

Table 11. General characteristics of Collyweston Slate (Woodward 1894).

		Thickness
Lincolnshire Limestone	Marly and oolitic limestones with occasional sandy beds	10 to 12 feet
	Sand with curious concretionary nodules and thin irregular slabs, that occur in undulating layers (in one place to the number of 37) and coalesce with oolitic and sandy stone at base (Top Sand)	3 feet
	Hard brown oolite; passing down into pale grey limestone and calcareous sandy stone (forming roof-bed in mine)	6 feet
Collyweston Slate*	'Slate' fine-grained calcareous sandstone	3 feet to 3 feet 3 ins
Lower Estuarine Series	Soft yellow calcareous sand and sandstone	

* Collyweston Slate is now included within the Lincolnshire Limestone Formation

> Collyweston Slate... These slates have been worked for upwards of 350 years at Collyweston and Easton near Stamford, Duddington, Medbourn, Kirby, and Dene Park, near Rockingham. (Woodward 1894, 170, 483)

As the formation which includes Collyweston Slate (Fig 49), the Lincolnshire Limestone is the most famous source of stone slates in the region. Delves were worked at Easton on the Hill, Collyweston, Duddington, Kirkby Lodge, near Deene and at Edith Weston but it has not been possible to confirm the geology of the beds worked at these sites for this report.

The lithological character of the beds at Collyweston is variable. Woodward (1894) described the general sequence (Table 11).

Collyweston Slate was mined from a very early date and split by frosting. It is worked by removing the Lower Estuarine Sand from below, temporarily propping the fissile rock or log until a large enough area has been undermined and then removing the props and allowing the log to fall. Sometimes the rock has to be wedged to persuade it to fall. It is then stored, usually underground, until there is a period of frost when it is brought out and allowed to split. Within the last century production difficulties have been experienced because of insufficiently severe, prolonged or frequent

Figure 49. A small Collyweston slate. Munsell 2.5Y, centimetre scale (© Terry Hughes).

Figure 50. Collyweston rock stacked after frosting (© Terry Hughes).

Figure 51. Once frosted the slates are clived with a hammer (© English Heritage Photo Library).

Figure 52. Dressing the edges. The method of dressing the edges of stone slates varies from region to region or even from delph to delph. It is only by encouraging or specifying the use of the specific local methods that the local distinctiveness of stone roofs can be conserved (© English Heritage Photo Library).

frosts. To overcome this, the Collyweston Stone Company has developed an artificial process. [24]

In 1875 Judd provided a first-hand account of the manufacture and use of Collyweston Slates. The version of Judd's report quoted here is from Woodward 1894 and includes some of his additional notes. Figures (references in square brackets) have been added for illustration.

> .... the Collyweston Slates have been dug over a considerable area, old pits being traceable from Wothorpe near Stamford to the western side of Collyweston, a distance of more than three miles. The valuable fissile character of the beds is merely a local accident; and in some locations the bed of stone has been followed and found to become non-fissile and in consequence worthless for roofing purposes. There is only a single bed of stone (the lowest limestone of the series) which is used for making roofing-slates. This varies greatly in thickness, being often not more than 6 inches thick, but sometimes swelling out to 18 inches, and in rare cases to 3 feet; while, not unfrequently, the bed is altogether absent and its place represented by sand (or sandstone). Rounded mammillated surfaces, like the 'pot-lids' of Stonesfield, abound in these beds.
>
> The slates are worked either in open quarries or by drifts (locally called 'fox-holes') carried for a great distance under ground, in which the men work by the light of candles. The upper beds of rock are removed by means of blasting, but the slate-rock itself cannot be thus worked, for though the blocks of slate-rock when so removed appear to be quite uninjured, yet, when weathered, they are found to be completely shivered and consequently rapidly fall into fragments. The slate-rock is therefore entirely quarried by means of wedges and picks, which, on account of the confined spaces in which they have to be used, are made single sided. The quarrying of the rock is facilitated by the very marked jointing of the beds, a set of

Figure 53. Making the fixing hole with a bill and helve (© English Heritage Photo Library).

master-joints traversing the rocks with a strike 40° W. of N. (magnetic), while another set of joints, less pronounced, intersect the beds nearly at right angles.

During the spring of the year the water in the pits rises so rapidly that it is impossible to get the slates out. The slates are usually dug during about six or eight weeks in December and January. The blocks of stone are laid out on the grass, preferably in a horizontal position. It is necessary that the water of the quarry shall not evaporate before the blocks are frosted, and they are constantly kept watered, if necessary, until as late as March. The weather most favourable to the production of the slates is a rapid succession of sharp frosts and thaws. If the blocks are once allowed to become dry they lose their fissile qualities, and are said to be 'stocked'. Such blocks are broken up for road-metal, for which they afford a very good material. The limestone beds above the slate-rock are burnt for lime

After the blocks are split, the slates are stacked on edge in circular piles or heaps [Fig 50]. Subsequently they are shaped, and again stacked on edge according to size.

The slates are cleaved at any time after they are frosted [Fig 51]. Three kinds of tools are used by the Collyweston slaters. The 'cliving hammer,' a heavy hammer with broad chisel-edge for splitting up the frosted blocks. The 'batting hammer' or 'dressing-hammer,' a lighter tool for trimming the surfaces of the slates and chipping them to the required form and size [Fig 52]. The 'bill and helve,' the former consisting of an old file sharpened and inserted into the latter in a very primitive manner. This tool is used for making the holes in the slates [Fig 53] for the passage of the wooden pegs, by means of which the slates are fastened to the rafters of the roof. These holes are made by resting the slate on the batting hammer and cutting the hole with the bill.

The slates are sold by the 'thousand,' which is a stack usually containing about 700 slates of various sizes, the larger ones being usually placed on the outside of the stack. The slates when sold on the spot fetch from 23s. to 45s. per thousand. Many of the Collyweston slaters accept contracts for slating, and go to various parts of England for the purpose of executing their contracts.

The land at Collyweston is generally held by slaters by copyhold, the slaters paying 6s. 8d. per pit' to the lord of the manor (a 'pit' is 16 square yards) with an extra charge of 1s. 6d. per pit to the measurer. A few workings are rented of the lord of the manor, the slaters paying 30s. per pit with an additional 1s. 6d. for the measurer. These payments are made every year at the annual 'slaters' feast' held in January.

The manner in which the slates are placed on the roof is as follows. The largest are laid on nearest the wall plate, and the size of the slates is made gradually to diminish in approaching the ridge. The ridge itself is covered by tiles of a yellowish white tint, made at Whittlesea, and harmonising well in colour with the slates themselves. The larger slates are, in the ordinary way, fixed to the rafters of the roof by means of wooden pegs driven through a hole in the upper part of each slate. But roofs are often covered with small slates which are fixed by mortar. ....

The slates of Collyweston are worked with more or less vigour at the present time [1889], although in many new houses built in the neighbourhood of the quarries, and at Stamford, brick and Welsh slates or red clay-tiles are employed, in place of the freestone and Collyweston Slate.

In colour the rock is a buff and blue-hearted stone, so that some of the slates are blue, others yellow, and many are parti-coloured. The pale coloured slates when put up, are said to darken on exposure. The slates are usually cemented as well as pegged on to the roofs, hence they do not fall away if cracked. The blocks that are raised from the open quarries and galleries are of irregular shape.

The slate-pits at Kirby are now almost entirely abandoned, and they are only occasionally worked near Dene Lodge. (Woodward 1894, 483–4)

Northampton Sand Formation: Duston Slates

The Northampton Sand Formation has been quarried or mined for stone slates since at least 1712 at Duston and Harleston Heath, to the west of Northampton, Pitsford and Weston Favell, the latter probably near surface rather than underground (Diana Sutherland, personal communication), to the north and east respectively. In 1870, the Duston Stonepit was described as 'very ancient and large some 40 feet deep, in which two beds of "white pendle" [25], each two or three feet thick occur between upper and lower sandstones' (Sharp 1870, 370) (Table 12). It is not known when they ceased working but by 1906 they were no longer used (Thompson 1906).

In the Duston area, several delves and mines were in existence, working a variety of building stones. The lower Pendle bed was called (incorrectly) Collyweston Slate by the quarrymen and was mined in a similar way (Sharp 1870, 370), although it is not clear whether it was a frosting slate. There is some confusion about the names and locations of the various workings but the correct situation is shown in Figure 54. At Pitsford and Weston Flavell, the slates were suitable for use straight from the ground without frost treatment (Morton 1712, 109). It seems that in the East Midlands only here and at Duston was the Northampton Sand Formation sufficiently sandy and fissile to be useful as slates (Diana Sutherland, personal communication).

Market

It has not been possible to make any assessment of the demand for slates other than Collyweston. In practice, if a roof of any type is to be repaired, either reclaimed stone slates (described as Collyweston no matter what their original source) or new Collyweston Slates will be used *de facto*.

A feel for the market can be gained from the condition survey which was carried out for the Peterborough Conservation Area Partnership Scheme which covers Collyweston Slate roofs in 14 villages (Christine Leveson, personal communication). A total of 69 roofs in need of repair were identified. All of these were eligible for grant support and the scheme very sensibly restricted the budget so as not to overwhelm the supply of new slates. Nevertheless, there has always been a demand in excess of

Table 12. Section of Old Duston stone-pit, giving quarrymen's terms.

1	white sand
2	brown soft sand with vertical plant markings (root perforations?)
3 The Roylands	a series of beds each from 6 to 9 inches in thickness, very variable, sometimes hard, in which condition it is 'best' building stone, and sometimes 'caly' or crumbling . These beds occur in two divisions, the building stone, of the upper being of a rich red brown colour, and the latter of a colder fawny-brown colour. Wood is frequently found and I obtained from these beds a slab ripple marked; sandy zones also occur in which the tests of shells are perfectly preserved.
4	orange sand with rounded cores of arenaceous limestone, the remains probably of the original bed after being subjected to the action of water charged with carbonic acid
5 White Pendle	coarsely granulated limestone, made up sometimes of oolitic grains in a matrix of calcareous cement, sometimes of crystalline angular particles with comminuted shells, more or less arenaceous in places, and containing *belemnites*, large in two beds, *Lima*, nov. sp. large *Hinnites abjectus*, etc Arenaceous and calcareous slaty beds, very like to and called by the pitmen 'Collyweston Slate'
6 The Yellow	building stone, consisting of six or seven beds of varying thickness, in two divisions, differing somewhat in tone of colour; these beds contain 'potlids' of ironstone, also *Cardium cognatum* etc
7 Best Brown Hard	building stone in three or four beds, a coarser stronger stone than that of the other beds, but of a rich brown colour: it contains few fossils
8 Rough Rag	a slightly calcareous sandstone, green-hearted, hard and durable, used for copings, gravestones and building: it contains *Ammonites murchisonae, A opalinous, Nautilus, Ceromya bajociana, Phalodomya fidicula, Cardium cognatum, Cucullaea* etc, and a characteristic zone of *Astarte elegans*
9 Hard Blue	a hard blue-hearted stone, the surfaces of joints and bedding brown from oxidation: it contains the same fossils as the last bed, no 8, excepting *Ammonites murchisonae* and the *Astarte elegans*
10	the presence of water prevents the working of stone in this pit to a lower depth; but in an old unused pit in an adjoining field the beds for about three feet lower are exposed; and these consist of cellular ironstone, having sometimes arenaceous, and sometimes ochreous cores

the budget. In the first three years 19 roofs were repaired, accounting for a grant allocation of £69,608 in a total re-roofing cost of £204,140. Because of the over-demand for help with re-roofing costs, the grant aid was reduced from 40% to 25%, but this did not reduce the number of applications. This scheme has been successful in conserving many roofs which might otherwise have been lost.

Outside the conservation areas, there is a substantial number of buildings with Collyweston roofs, some listed but many not. There is, therefore, sufficient demand to support a commercial manufacturing industry at a modest level of output.

Production

Effectively, there is no commercial production of any of the region's slates, although some roofing contractors, to their credit, have continued to produce sufficient

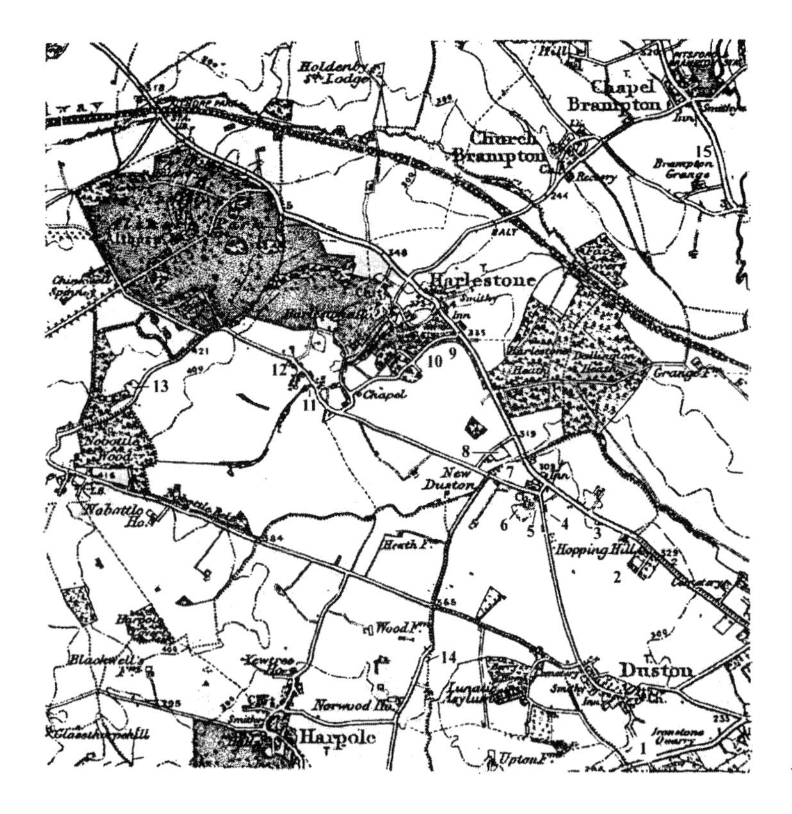

Figure 54. Stone workings in the Duston area (annotated by Richardson 1904; names in bold are stone slate delves). 1 Duston Limestone; 2 Limekiln Quarry; 3 Hopping Hill Brickworks, not annotated; 5 Field Pit; ***6 Top Pit New Duston Old Duston Pit (Sharp);*** *7* ***Tennant's Quarry = Sharp's Old Slate Quarry Close, Northampton Sand Formation;*** *8 Richardson's Slate Pit Plantation, Northampton Sand formation; 9 Cotter Quarry, Harlestone; 10 Larger Quarry; 11, 12 and 13 were not annotated; 14 and 15 Harpole Sandpit. Quarry 9 is misplaced; it should be 0.4 km further south.*

Collyweston Slates for their own needs for many years, taking log from an aggregate quarry or by operating a mine at a very low level of output. That there is no commercial production is very disappointing, especially in view of the fact that sources of slate log are well known in the area and that the Collyweston Stone Slaters Trust has been in existence for 17 years specifically to solve the problem of supply. It appears that the situation requires a strong commercial drive which has been lacking in the past, and that the best hope for solving the problem would be the introduction of some entrepreneurial initiative, if necessary from outside the region. Happily in 2001 there were signs that this was happening. Details of manufacturers can be found in Annex A.

It has been a tradition in this region for roofers to mine and manufacture their own slates. While this has certain advantages in terms of quality control, it has not succeeded in securing an adequate supply of slates to the market in general. It would, therefore, be unwise to support this as a strategy for production in the future. What the market needs is a proper commercial operation which will provide adequate amounts of slate with a short lead time and at a reasonable price.

Before any commercial production can start, an artificial freezing process is needed. This has now been developed. Without the artificial process the efforts that have gone into other aspects of conserving roofs in this region were always doomed to be less effective than they could otherwise have been because, without the artificial process, any substantial production was reliant on satisfactory periods of natural frost, a situation which often failed to materialize.

Other sources

Other than the small amount of new slate, the main source of Collyweston Slates for re-roofing and new build is cannibalization of other roofs. There is no evidence of similar slates (that is, Stonesfields) being imported from outside the region.

Summary

In the past, the East Midlands used stone slates from a variety of sources. Nowadays, only the Collyweston type is available as new and this only in small quantities.

Collyweston has been better placed to solve the problem of supply than almost any other stone slate region in England and the fact that production over the last ten years has been at best spasmodic is very disappointing. There seems to be no technical, market or economic reason why this should be the case and therefore what is needed is a strong commercial drive.

THE SOUTH EAST OF ENGLAND

Only the Horsham Stone is well known as a source of stone slates in this region but it appears that this stone may have been exploited for roofing over a much wider area than is commonly assumed. Recently two other stone slates have come to light which are thought to be of Wealden age but not from the Horsham stone.

Horsham

Horsham Stone is a group of sandstone beds in the Weald Clay. It is the principal stone of this belt and has its greatest exposure around Horsham, where it has been used for building, roofing and flagging since Roman times (Fig 55). The earliest use would have been very localized, each building obtaining what was needed from adjacent fields, but as transportation improved it came to be used throughout Sussex, parts of Kent and even in London. The last active quarry appears to have been at Nowhurst in 1939. At that time stone slates were considered to be uneconomic in their own right but were produced 'at a nice price' in the course of delving for paving and roadstone.

> Nowhurst has been quarried for several hundreds of years but the roofing stone comes in patches. We find a stratum about three feet down with now and again a hard patch in it. This stratum varies in depth from two feet to five feet and it is from this that the roofing stone comes. The base of the stone is sandstone but for some reason there is lime in the roofing stone patches which hardens it into a granity stone. You can't split any stone for roofing. The natural cleave has to be there. If the stone has a tight cleave we let it stand in the wind and frost for some months, and the weather helps to split it perfectly. (Dunkerton, 1945, 213)

Figure 55. Typically made of large, heavy stone slates, Horsham roofs are simple and robust (© English Heritage).

Figure 56. Today, Horsham stones are frequently edge-bedded and pointed with mortar (© Lisa Brooks).

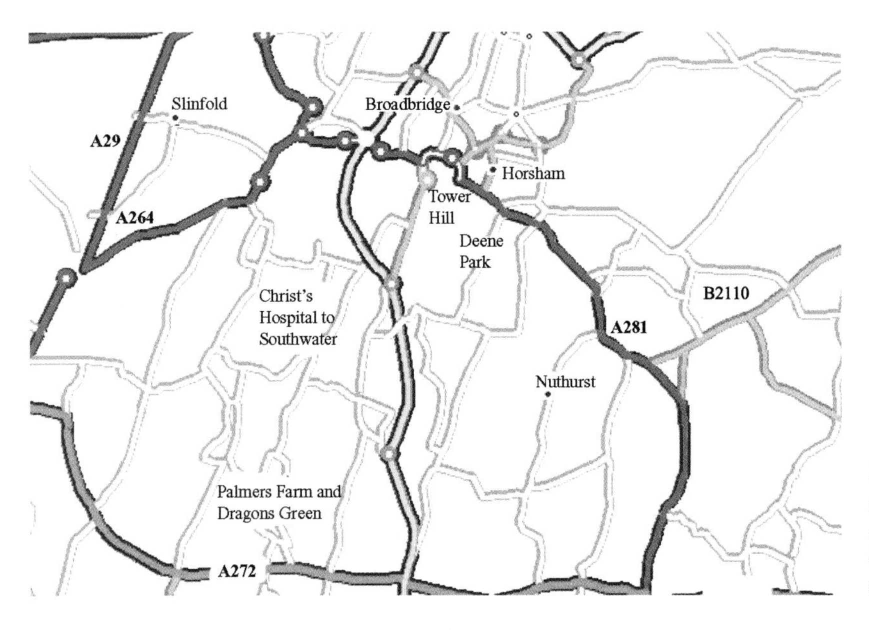

Figure 57. Some locations of old Horsham delves (Reproduced by permission of Ordnance Survey mapping on behalf of the Controller of Her Majesty's Stationery Office © Crown copyright GD0385G/03/01).

Even before 1939 there had been difficulties obtaining stone slates to repair roofs. As a consequence a system of slating developed to make the available slates go further. Essentially this changed the roofing from a double-lap into a single-lap system which relied on shadow slates and mortar to make the roof weather proof (see 'The construction and style of stone slate roofs' below) (Fig 56).

Geology

The Horsham Stone is described in the Geological Survey Memoir for the district (Gallois and Worsam 1993) (Table 13). It can be traced as a single seam from Crawley to Horsham, and south-east from there as two seams until it passes into the Brighton district. Within these seams the stone is generally present as distinct beds separated by thin clays. They may be more or less calcareous and the fissility varies. The upper bed is often weathered and decalcified and is not suitable for building purposes. The middle bed is thinly-fissile, frequently ripple-surfaced, and suitable for roofing and flagging. Below this, the stone is massive and has commonly been worked as an aggregate.

A substantial stone industry developed to the east and south of Horsham, with large quarries at Christ's Hospital School, Stammerham, east of Itchingfield, around Nowhurst, and in the Tower Hill to Deene Park area. Beneath the Horsham Stone is clay-ironstone which has been worked as extensions to stone quarrying. Equally, the Horsham stone was often worked to get to the ironstone. Known quarry sites (Gallois and Worsam 1993, 80–2) are shown in Figure 57 and listed in Table 14. Although many of the historic sites are now built over or otherwise sterilized, core drillings and exposures within the area indicate that the fissile rock is still accessible at several locations. A typical section, reproduced as Table 15, was recorded by Lyell in Stammerham Quarry (Gallois and Worsam 1993, 81).

Other stone slates in south-east England

Two other stone slates have come to light. One, from a roof between Guildford and Oxted, might have been presumed to be Horsham. However, it is geologically and visually very different. It is a fine-grained, argillaceous,

Table 13. The position of Horsham Stone in the Lower Cretaceous.

Age	Rock Unit		
Lower Cretaceous	Weald Clay	Upper division Lower division	Horsham Stone
	Hastings Group		Tunbridge Wells Sand Wadhurst Formation Ashdown Formation
	Durlston Formation		
Upper Jurassic			

Table 14. Historical Horsham Stone quarries and other exposures.

Location	Site	Grid reference
Warnham	Ends Place to Warnham Court Road section	TQ151333
	Horsham Corner Wood	TQ148330
	The Pits (Clay-ironstone)	
	Wasp Pit (Clay-ironstone)	TQ142323
Slinfold	Theale Copse	TQ127319
	Birch Copse	TQ127318
	Quarry Field	
	exposure in the bank of the River Arun 400 m (44 yards) east of Hill House	
	diggings between Rapkyns and Brookhurst including Nowhurst Quarry	TQ135325
Broadbridge	High Wood. Clay Ironstone pits	TQ147300
	stone pits in an abandoned fold of the River Arun	TQ143297
	stammerham Quarry (Location uncertain)	TQ146292
	roadside quarry	TQ154287
	sparrow Copse	TQ151294
Christ's Hospital	on the dip slope 0.5 to 1 km (54.5 yards) south	
	Two Mile Ash	TQ149271
	exposure	TQ157274
Palmers Farm to Dragons Green	The Delph	TQ155246
	diggings in a copse	TQ142241
Tower Hill and Deene Park	on the dip slope	
	Chesworth Manor (Brooks)	
Nuthurst	north-east of Kites Copse	
	exposure at The Gill	TQ182272
	exposure	TQ177260
	Harriot's Hill Ironstone pits? 400 m south of Nuthurst Church	TQ180257

Table 15. Section in Stammerham Quarry (Lyell)

Horizon	Thickness in feet
Vegetable mould (soil)	1.5
Stiff clay and loam	9
Compact calciferous sandstone with ripple marks on the upper surface (rough causeway)	4 in
The same rock but more indurated in two layers 0.1 and 0.3 m (4 in and 1 foot) (scrub stone; road material)	1.34
Ferruginous sandstone	1
Blue soapy marl	1.5
Ferruginous sandstone	1
Hard calcareous sandstone (ground pinning-stone)	1
Compact calciferous sandstone with ripple marks (paving)	2
Marl	4
Stone in slabs, reached by drilling	

Figure 58. Horsham stone at Lower Broadbridge Farm (Terry Hughes).

Figure 59. Horsham stone from Lower Broadbridge Farm in production trials by the Completely Stoned Company (© Terry Hughes).

lithic greywacke with a gritty surface and is yellowish grey in colour (5Y 7.5/1 on the Munsell colour scale) (Jefferson 1998). At present, its origin is unknown. It has been provisionally placed within the Wealden formation but is not a Horsham stone and, if it proves to be from a local source, is a rare example of a stone slate of an unusual rock type. [26] The other is known from Charlwood near Gatwick. It is believed to be a Cyrena limestone. These can form layers up to 50 mm (2 in) thick within the mudstones and consist of closely packed shells of *Filosina gregaria*. They occur within the same sequence as the Small Paludina limestone of the Weald Clay, well known as a masonry stone (Sussex Marble, Bethersden Marble, Petworth Marble in Surrey and Kent) (Graham Lott personal communication). It sometimes turns up on Horsham roofs (David Ansell, personal communication).

Market

The market for Horsham roofing is quite large and reasonably constant. Local opinion is that there would be an even larger demand if supplies of new stone could be obtained. The implication is that cost would not be a serious hindrance. However, the market will be smaller than it might be, if the use of the single-lap slating system were to be abandoned (see 'The construction and style of stone slate roofs - Horsham roofs', below). Objections to such a change could be expected, partly on cost grounds and partly due to increased loading. Also, many roofs were originally built using single lapping and would presumably be conserved in this style. That aside, it is technically desirable to return to double lapping, if only because some 'single-lap' roofs have failed or needed premature renovation.

Production

Horsham stone slates are not delved currently but may become available soon. In 1996, trial excavations were carried out at Lower Broadbridge Farm (Fig 58). A quantity of roofing and flagging was easily produced from rock close to the surface (Fig 59). This was assessed for English Heritage and found to be a compact siliceous limestone [27] of extremely low porosity, suitable as a source for tilestone, flagstones and walling stone.

There appears to be an adequate reserve to satisfy demand. Planning permission has been granted, and the landowner and a slate manufacturer are in discussions prior to starting operations. Initially this might be production to order so any potential purchaser would need to allow a sufficient lead time for the delving and processing to take place. The alternative of supplying stone slates from stock might need to be based on full exploitation of the reserve and therefore a market for flagging and walling would need to be developed. The prospects for such products are good.

Other sources

The only other source of Horsham stone slates in the region is reclaimed material. This is a long-established practice, probably through most of the last century. So difficult has the supply situation become that the technique, similar to single-lap slating and described below, has been developed to stretch the limited amount of slate.

Horsham stone slates appear to be sufficiently distinctive in colour and texture to render other British sandstones inappropriate as substitutes.

Summary

The problem of inadequate supply of new stone slates appears to have existed in the Horsham area longer than any other part of the country. The fact that so many roofs still exist is a tribute to the ingenuity of local roofers in devising methods of making limited slates go a long way. Unfortunately, some of these practices conflict with technical efficiency. The issues of just what is acceptable practice and the details which should be applied, need to be resolved.

Effective as the local methods of reslating have been in preserving stone roofs, without new stone Horsham roofing will eventually disappear. The initiative between Lower Broadbridge Farm and a slate manufacturer cited above should be supported. If it were to fail for lack of support from the market and conservation bodies there is little likelihood that anyone else would take up the challenge.

THE WELSH MARCHES AND BRISTOL

> Since the earliest days of the history of the science of geology the Welsh Borderland has attracted the attention of geologists by the great variety and interest of its formations, for in no other area can the sequence of the Palaeozoic rocks be seen to such advantage and in such a comparatively small district. (Earp and Hains 1971, 1)

Unlike other stone slate regions of England, the Welsh Marches have not been defined by their geology for this study. Geographically, the stone slate usage extends from the Bristol Channel to Shropshire. It may be defined to the west by the Welsh border, and to the east roughly by a line due north from Gloucester. For convenience, the Pennant stone of the Bristol region has been included with the Pennant of the Forest of Dean and South Wales.

The region cuts across the grain of the geological succession from the Ordovician to the Jurassic, a period of about 300 million years. This makes for interesting roofs but difficult research. Because the geology changes rapidly over short distances, and because so many different stones have been used for roofing, this section is inevitably incomplete. There is no doubt that many small delves, hidden away in forests, in deep narrow valleys or on remote hill tops, still remain to be discovered. Similarly, there are many sources of stone slates from the same formations across the border in Wales. These have not been included in this report but they should not be overlooked when searching for potential new supplies. From Llandeilo in South Wales to Ludlow the lowest part of the Silurian Přídolí Series consists of micaceous flaggy sandstones, known as the Tilestone Formation. (Today it is more correctly the Long Quarry Beds in the south and the Downton Castle Sandstone Formation in the north.) Its use is differentiated by the initial capital in 'Tilestone', otherwise used to describe any stone slate.

An example of the difficulty in determining the historical provenance of stone slate roofs in the region is provided by Ashleworth Tithe Barn, north of Gloucester, which was to be re-roofed during 2001 (Figs 60 and 61). The barn is thought to have been built between 1481 and 1515 but the roof was reslated in 1885 and again in the 1940s. In 1942, Harold Trew, a Gloucester architect, reported that the roof was covered with 'Cotswold stone tiles' (unpublished letter to the Society for the Protection of Ancient Buildings, 1942). In fact, the roof today includes red and grey-green Old Red Sandstone and Forest Marble slates but it is not possible to determine which is the earliest or even if they were both installed at the same time. The Old Red Sandstone is only six or seven miles away at Newent to the west or the same stone could have been brought up the Severn from the Forest of Dean area. The Forest Marble would have come overland from the Cotswold Hills fifteen miles to the east.

Figure 60. Ashleworth Tithe barn: probably as the result of successive renewals, this roof is a mixture of red and grey-green micaceous sandstones of Devonian age and Forest Marble, probably from the Cotswold Hills. The source of the sandstones could have been Dymock or the Forest of Dean (© Terry Hughes).

Figure 61. A mixture of Devonian sandstones and calcareous Cotswold stone slates on Ashleworth Tithe barn. The acidic sandstones are covered in green algae whereas the Cotswold stone supports grey lichens (© Terry Hughes).

To further complicate matters Ashleworth actually stands on rocks of Jurassic-Lower Lias age and field walls in the vicinity include thin stone which would be suitable, and therefore might have been used in the past, for roofing (Fig 62). Stone for field walls was never carried far. Indeed, stone walls often only exist because they are a 'waste product' of a nearby quarry which was primarily worked for more valuable products such as the Lias masonry of Ashleworth barn. The convenience of a nearby supply of Lias roofing might well have been the overriding factor in deciding what was used on the roof originally. (Sources of lias stone are discussed below.)

Figure 62. A thinly bedded Lias Limestone from Ashleworth near Gloucester. This stone might have been used for roofing in the past. Munsell 2.5Y (© Terry Hughes).

So, at present, there is a mixture of stone slates and an indeterminate history of other stones from an earlier date. In similar situations elsewhere, commentators have added further confusion by incorrectly identifying sandstones as limestones and vice versa, because they have incomplete knowledge of all the possible, local options.

Geology

The sources of stone slates in this region are listed in Table 16. The geology of the Silurian-Devonian boundary has been the source of much debate. In particular, in the past there has been uncertainty about where to place the 'boundary' formation, the Downton Series. It is now placed at the top of the Silurian and more correctly named the Přídolí Series, so care is needed when using older texts and maps. (Although the Downton Series has been described as part of the Devonian in geological texts for much of the twentieth century, in earlier descriptions, by Murchison, for example, it is correctly, described as Silurian.)

Ordovician

> If farm buildings were to be covered upon the roof with boards and afterwards with slates (which is the custom in London), they would be more durable, and not subject to be injured by the necessary work of moving hay and corn in the lofts, or barns, and which is committed by ignorant servants. The boards may be of deal, and three-fourths of an inch thick. The concluding observation removes an objection to the use of slates in preference to thatch. In the SW district of Shropshire, there are some good quarries of stoneslate, which though they require stronger timbers than the blue slate, are a valuable cover when at no great distance; and they would be much firmer when put on, if the ridge was covered with some crests instead of a turf or no crest at all. Tile glazed crests may sometimes chip by frost. Any cover is preferable, in look and duration, to the common clay tiles of this county. (Plymley 1803, 106)

In the north of the Marches a variety of Ordovician rocks have been exploited for roofing (Figs 63 and 64). These

Table 16. Sources of stone-slates in the Welsh Marches.

	Age		Rock unit	
Jurassic	Lower		Lower Lias?	
Permo-triassic	New Red Sandstone	Triassic	White Grinshill Sandstone - Grinshill Flagstone	
Carboniferous	Silesian		Westphalian D including the Pennant and Supra-Pennant Formations	
	Dinantian		largely absent	
Devonian	Upper Devonian	'Upper Old Red Sandstone'	not present in this area	
	Mid Devonian	'Lower Old Red Sandstone'		
	Lower Devonian		Breconian Ditton Series	Brownstone group St Maughan's group
Silurian	Prídolí		Downtonian	Downton Castle Sandstones (The Tilestones)
	Ludlow			
	Wenlock			
	Llandovery			
Ordovician	Ashgill			
	Caradoc		Cheney Longville Flags, *alternata* Limestone, Chatwall Sandstone, Chatwall Flags?, Hoar Edge Grits	
	Llandeilo			
	Llanvirn		contains the Corndon Hill quartz-dolorite intrusion	
	Arenig			
	Tremadoc			

Figure 63. Wilderhope Manor. The near slope is an unidentified local sandstone; the rear slope is Harnage. The different lichens, yellow growing on the acid sandstone and grey on the calcareous Harnage, can be seen (© Terry Hughes).

include the Cheney Longville Flags, the *alternata* Limestone, the Hoar Edge Grit and, possibly, the Chatwall Sandstone and Chatwall Flags. Greig *et al* 1968 includes a useful table of the quarries in the region. Table 17 is an extract of those in relevant formations.

The horizons named in Table 17 are those of the northern part of south Shropshire. For those with different names in the southern part the equivalent horizons are Coston Formation (Hoar Edge Grit), Glenburrell Formation (Chatwall Flags) and Horderley Sandstone Formation (Chatwall Sandstone).

South Shropshire district

> In the same region (Shrewsbury) a greenish-grey micaceous sandstone has been used to produce smooth-faced, small and heavy tile-stones known as 'Cheyney Longville

Figure 64. Fine-grained sandstone from Wilderhope Manor. Munsell Gley, centimetre scale (© Terry Hughes).

> slabs'. They have been found during recent archaeological excavations of medieval houses in Shrewsbury and may still be seen on a few of the roofs of buildings at Bishops Castle and Whitcott Keysett. (Carver 1983, 32)

Within the area roughly bounded by Ludlow, Knighton, Bishop's Castle and Church Stretton there are many isolated stone slate roofs and references to their use. Unfortunately, because of the variety of the geological formations, the visual similarity of many of the stone slates to each other and movement of slates outside their natural district in recent years it is difficult to be certain which rocks are on which roofs today. To resolve this it would be necessary to carry out petrographic examinations of

Table 17. Quarries in formations which have produced stone slates south of Shrewsbury (after Greig et al *1968).*

Quarry	Grid reference	Hoar Edge Grits	Chatwall Flags	Chatwall Sandstone	Cheney Longville Flags	
					Lower	Upper
Chatwall Hall	SO5137 9758		+			
Cheney Longville	SO4173 8584			+		
	SO4120 8512			+		
	SO4154 8593			+		
	SO4207 8546				+	
Cheney Longville Old	SO4101 8524			+		
Coston	SO3860 8008	+				
Cwm Head	SO4188 8797	+				
Enchmarsh Old	SO4978 9622			+		
Hope Bowdler	SO4711 9174		+			
Hope Bowdler	SO4772 9182			+		
Soudley	SO4772 9182				+	
Horderley	SO4118 8614	+				
Horderley	SO4136 8658	+				
Marshbrook	SO4446 8901					+
Marshbrook Old	SO4406 8976					+
Shipton Old	SO5629 9174					
Sibdon Carwood - Longlane	SO4127 8422			+		
Willstone Old	SO4898 9562			+	+	
Wittingslow Old	SO4302 8834					+
Woolston Old	SO4222 8734			+		

Figure 65. Cheney Longville flag roof at Clun near Whitcott Keyset (© Terry Hughes).

samples taken from roofs. However, originally they would all have been used locally.

The formations listed above which have been associated with stone slate production lie approximately parallel to each other along a line trending roughly south-west to north-east with the older, Hoar Edge Grit further to the north-west.

Chatwall Sandstone

Greig *et al* have described this rock as flaggy in several delves and markedly so around Wart Hill (Greig *et al* 1968, 126–7). At Longville Common the rock is Chatwall Sandstone but is mostly hidden by trees. The few exposures do not show any fissile rock. However, two small delves at Long Lane north of Sibdon Carwood do contain fissile Chatwall Sandstone. Martin (1986) has referred to tilestone beds in the top few feet of the Soudley Sandstone (the local name for the Chatwall Sandstone). Soudley quarry (SO 477918) shows little fissile rock, but what there is, is high up. The Chatwall Flags (also known as Soudley Sandstone) are reputed to have been used for roofing but it was actually obtained from shallow pits in the *alternata* limestone which overlies the Chatwall Flags in the fields above the present quarry. The beds are packed with the flat fossil shells of a brachiopod, *Heterorthis alternata*, which gives it its fissile quality (Andrew Jenkinson, personal communication). There are still a few *alternata* roofs in the Hope Bowdler/Church Stretton area.

Cheney Longville flags

La Touche recorded that the 'Longville Flags have been used for roofing' but the quarries were not detailed (La Touche 1923, 62) (Fig 65). Two quarries in the Cheney Longville Flags are recorded by Greig *et al*: 'Khaki flags and shales have been worked in two old quarries in the plantation 1270 yd W 4E N and 1180 yd W of Wistanstow Church. About 20ft of beds are seen in the more westerly quarry [SO 420857]' (Greig *et al* 1968, 131). The flags were extensively delved for flooring and roofing between Minton on the flanks of the Longmynd and Cheney Longville village. There are several small delves in the outcrops around Marshbrook and another near Winstanstow at SO423860 (Andrew Jenkinson, personal communication).

Figure 66. Harnage slate from the Hoar Edge Grit (© Terry Hughes).

Hoar Edge Grit

Grieg *et al* recorded several quarries north-east of Cheney Longville containing flaggy rock, for example SO 412861, SO 412862, but there are no specific references to their use as roofing (Grieg *et al* 1968). However, since most of the references to quarries and other exposures note the presence of brachiopod shells, which often form the plane of parting for these stone slates, it is possible that they could have been produced in any of these quarries. Clifton Taylor stated that the stone slates of Stokesay Castle were obtained from the Hoar Edge Grit at Hoar Edge (SO 975974) without giving any evidence for this conclusion (Clifton Taylor 1983, 224). He also implied that they were thin, [28] which is certainly not a description of the Harnage (Hoar Edge) stone slates from near Acton Burnell.

The Hoar Edge Grit has been an important local source of stone slates in the area to the south of Shrewsbury around Acton Burnell (Fig 66). The extent of their use was reviewed by Lawson and the Shropshire Archaeological Society has listed the 18 remaining examples (M Moran, personal communication). One, the Church of St Michael and All Angels, Pitchford, was re-roofed during 1999 using newly delved slates from near Acton Burnell (see Wood and Hughes, this volume). In an extensive field survey the ridges of Park Wood and Grange Hill (SJ 535006 to SJ 570019) were found to have been worked as a series of small, shallow pits and more substantial delves in a sandy and shelly facies of the Hoar Edge Grit. Also, at Bull Farm, a large quarry has worked

Figure 67. A red roof at Michaelchurch Escley. In the Olchon Valley stone for the buildings and roofs was quarried on site so the roof colour depends on whether the grey-green or red stone was nearest to hand (© Terry Hughes).

Figure 68. St Mary's Church, Craswall: a grey-green roof (© Terry Hughes).

similar stone low down on Grange Hill (SJ 558015). To the north and the south-west of these two hills the Hoar Edge Grit can be followed in a series of exposures and quarries but none of these contain the same fissile rock. The results of the survey are recorded in Hughes and Jefferson (1998).

The stone slates produced during 1998 were shelly sandstones, with uneven surfaces and up to two inches thick (50 mm), ranging in size from 28 to 10 inches long (710 to 250 mm). [29] When freshly delved, although they appeared to be fragile, they were remarkably strong and hardened further after drying. They were squared and sized with a diamond saw, then split by hand and the edges dressed with a hammer (see Fig 3). The planes of parting are easily seen because of the orientation of the shells, especially in a sawn edge.

Devonian and Silurian

As a stratigraphical name, the Old Red Sandstone is obsolete but it is still useful in grouping together stone slates which are of Silurian and Devonian age. The Lower Old Red Sandstone, which spans the lowest part of the Devonian and the Přídolí, has been a source of roofing stone of Dittonian and Downtonian age. Included within this group, at the base of the Přídolí, is the Downton Castle Sandstone Formation, another source of stone roofing. It was formerly known as the Tilestones Formation and extends across South Wales to Llandeilo north of Swansea.

Stone slates have been produced all over the region, often in unrecorded, small delves. Given the complexity of their production and the fact that the geology of the Old Red Sandstone changes over such small distances, it has proved impossible to decide precisely which part of the succession has been exploited for roofing at some locations. Where a source is recorded without geological information (or is being sought for future production) petrographic investigation and field mapping will be required to determine the rock type and its suitability.

Predominantly, these stone slates are either grey-green, red or a reddish purple, fine to medium-grained, micaceous sandstones (Figs 67 and 68). They are more or less calcareous, a property which has been implicated in their variable durability. The fineness of the lamination is also variable and slate sizes tend to be smaller than the sandstones of the other regions, frequently no larger than 24 inches long (610 mm) and down to 8 inches (200 mm). Today, many of the roofs in the region include a mixture of the colours. Given the close proximity of the different colours and the small scale of the delves this is probably an original feature.

Dittonian and Dowtonian

Forest of Dean

The Old Red Sandstone surrounding the Forest was exploited at many quarries. Again, specific stone slate delves are not known. Any slates produced on the east side of the Forest would have had access to markets along the River Severn and on the west along the valley of the River Wye.

The Olchon Valley to the Golden Valley

Within this area there are many small, disused delves, most of which probably produced stone slates to some extent. Known sources include Coed Major (Figs 69 and 70) and Grigland and Pennsylvanni delves (Fig 71) at Llanveynoe, Pikes Farm (Alan Haseldine, personal communication) and Lower House Farm (ibid). It appears that the original roof of Dore Abbey was obtained from land on the border of Bredwardine and Moccas (see 'History' above).

North of Hereford

Many stone roofs can be found along the valley of the River Lugg and in the surrounding countryside. They include both grey-green, red and purple slates, often mixed on the same roof (Figs 72 and 73). Two sources, involving several delves, at Dinmore and Garnons Hills, have been recorded in detail.

> Dinmore Hill. The Dittonian sandstones may be seen in quarries at Queen's Wood and and also at Howe Wood. One of the Queen's Wood quarries is very large though very disappointing from a geological point of view … The

Figure 69. Delving grey-green slates in the late 1990s near Craswall, Herefordshire (© Terry Hughes).

Figure 70. Grey-green, micaceous sandstone from Llanveynoe in the Olchon Valley. Munsell 2.5Y (© Terry Hughes).

Figure 71. Trials in the old Grigland Quarry near Michaelchurch Escley. This and the adjacent Pennsylvanni Quarry are producing red stone slates (© Terry Hughes).

Figure 72. Red, micaceous sandstone from Morton on Lugg. Munsell 7.5YR (© Terry Hughes).

Figure 73. Grey-green, micaceous sandstone from Morton on Lugg. Munsell 2.5Y (© Terry Hughes).

> rock is a flaggy purple sandstone which splits so well as to make it a tilestone and probably this was the original purpose of the quarry. The other quarry [on the west side of Queen's Wood] shows massive sandstone with tilestones above.
>
> Near Kipperknoll, on the west side of the Burgehope wood, is a small overgrown quarry where the highest beds of Dinmore may be seen. They consist of well bedded brown micaceous sandstone forming tilestones with a band of very hard close textured cornstone about six inches thick. (Clarke 1951, 228)

> Garnons Hill. The rocks of which the hill is made are bands of sandstone of varying shades from pink to pale green, and from dark purple to brown. The sandstones are sometimes massive and extremely hard, sometimes soft and flaggy. Some of these split so easily as to form tilestones.
>
> Roofing Tiles - Some of the sandstones on the hill split naturally into tiles about an inch or a little more in thickness and these were once quarried extensively for roofing. Most of the older buildings in the neighbourhood are still roofed with these tiles, eg Mansell Gamage church and Byford Court, and originally probably all were. The disadvantage is that they are extremely heavy and need a very substantial structure beneath to carry them. (Clarke 1950, 102) [30]

These are only two examples of a much more extensive delving in the region. At Dinmore Manor the stone roofs of the chapel and hall, the latter now largely replaced, is believed to have come from a delph in Bridge Wood on the south side of the valley although this was not mentioned by Clarke in his description of the Dinmore delves. This is typical of the local, short-term delving which must have been the source of the stone roofs on many of the buildings in the region such as Lower Brockhampton Hall, near Bromyard.

East of Hereford

> The hard sandstone layers in the Rushall Formation were pitted at Hagley and along the eastern side of Shucknall Hill, presumably for local building stone. Murchison reported [1854, 135] that the Upper Ludlow Beds were much quarried for wallstone, but were very prone to weathering. They may also have been used as a flagstone or for roofing tiles. Many small pits occur in them at Shucknall Hill and around Stoke Edith, and were worked for rough building stone and road aggregate. [Phillips 1848, 74]

In a brief field survey during 1998 no stone roofs could be found here or at Yarkhill, but the disused delph on Shucknell Hill contains fissile rock. The absence of stone roofs might be explained by the comment that the Upper Ludlow Beds were very prone to weathering - they may have been replaced long ago (Murchison 1854, 135).

In the same area, Howe (1910, 13) mentioned that 'a grey, red and brown micaceous sandstone delved near Downton Castle, Herefordshire and Daymock near Ledbury is reputed to have been used as tile-stones' but the geology was not specified (Daymock is now Dymock). The use of stone slates further south at Ashleworth has been described above.

Figure 74. Micaceous sandstone at Willersley, Herefordshire (© Terry Hughes).

Figure 75. Old Red Sandstone roof at Winforton, Herefordshire (© Terry Hughes).

North of Hay-on-Wye

> micaceous tiles for roofing from Clayrow Hill in the red sandstone tract. (Davis 1815, 146)

In spite of difficulties in finding replacement stone slates, the area between Hay-on-Wye, Presteigne and Leominster still has a rich stone roofscape. Eardisley and Hay are outstanding examples, although, even here, many stone roofs have been lost in recent years. It appears that the Old Red Sandstone and the stone slates from the 'Tilestones Formation' of the Přídolí have both been used for roofing in this area (Figs 74 and 75).

Ludlow

Howe recorded that 'the thicker beds of a soft micaceous flagstone quarried in the vicinity of Ludlow and known as "pendle" by the local quarrymen are reputed to have been used for roofing purposes' (Howe 1910, 142). The specific geology has not been determined because of the imprecise location but this too may refer to the Downton Castle source mentioned above and described by Murchison: 'The upper portion of the Ludlow formation, or capping of the bone-bed, is composed of light-cloured, thin-bedded, slightly micaceous sandstones, in which quarries are opened out near Downton Castle on

Figure 76. A tilestone roof near Llandeilo, Dyfed (© Terry Hughes).

the Teme. ... The uppermost layers of the whole [Silurian] system, and which form a transition into the Old Red sandstone, consist of tilestones and sandstones, occasionally reddish' (Murchison 1854, 138).

Presteigne

There are many old quarries in the Silurian/Přídolí around Presteigne which were undoubtedly sources of stone slates in the past. Near the village of Evenjobb there is a fairly large delph which has fissile rock similar to the stone slates on Clunbury Church (John Wheatley, personal communication). The location indicates that it is Ludlovian in age.

Near Coombe Moor at SO 366630–368630 there is a large exposure with thinly fissile rock and lots of 'roofing' lying at its foot and in the scree below. At SO 362623 alongside the road to Stansbatch there is a small delph with thinly bedded rock near the surface and lower in the exposure. There are several quarries on the Forestry Commission land to the north of the road. A sample from Coombe Moor has been tentatively identified as being from the Ludlovian Series of the Silurian (David Jefferson, personal communication).

Tilestone Formation

> The Red Soil Tract ... Varieties. Though the rocks of this tract exhibit few or no anomalies, they have nevertheless three varieties or families. The middle strata consist of micaceous schistus: the thinner sort are converted into roofing tiles, and those from two to six inches thickness into flags, mile-stones, &c. The interior of the tile or flag is a compact sand-stone, of varieties of colours, some brown, some greenish, and others grey. What occasions the cleft, some half an inch, and others several inches distance from each other, is a thickly bespangled bed of mica. Wherever this bed of mica occurs, there is a cleft practicable, and in no other part: the transverse fracture exhibits no mica. Milestones on the Llanbedr (Lampeter) road, are from these micaceous quarries at Pont y Llechau. The micaceous pennant-stone of the coal tract, and the fire-stone of Bettws in the red soil tract on the northern border of Radnorshire, are of this family. (Davis 1815, 40) [31]

This part of the Lower Old Red Sandstone was originally given the name 'Tilestones' by Murchison at Llandeilo in West Glamorgan, where roofing (Fig 76) was produced in a series of long, narrow quarries. He describes the route of these 'finely laminated, hard, reddish or green, micaceous, quartzose sandstones' in a north-easterly direction across South Wales and into Herefordshire and Shropshire.

> This lower division of the Old Red System, though of much smaller dimensions than the overlying formations, has very marked characters both in structure and fossil contents, and is very clearly defined by occupying a position in which it passes upwards into the cornstone and marls, and downwards into the Silurian rocks. In this relation it has been already alluded to at Pont-ar-lleche (bridge on the tiles [32]), near Llangadock in Caermarthenshire, from whence it is seen to run in a nearly rectilinear course, from the Tri-chrug on the south-west, to near Builth (SO 044507) on the north-east, occupying the loftiest part of the escarpments of the wild tracts of Mynidd bwlch-y-groes and Mynidd Epynt, at heights of fifteen hundred and sixteen hundred feet. In this range, the tilestones are extensively quarried, and the strata, which are inclined at seventy and eighty degrees near Pont-ar-lleche, diminish to forty and forty-five degrees at the north-eastern end of the Mynidd Epynt, the dip being invariably to the south-east. After a great flexure on the Wye, to the east of Builth, the tilestones are again found in similar relations overlapping the Silurian rocks in the Begwm and Clyro Hills, Radnorshire, and extending thence to Kington in Herefordshire; in which part of their range they are much less inclined. Throughout their course from Caermarthenshire to Kington, the distinguishing beds are finely laminated, hard, reddish or green, micaceous, quartzose sandstones, which split into tiles. Although the greenish colours prevail, these beds are usually associated with reddish shale, and the decomposition of the mass uniformly produces a red soil, by which character alone the outline of the division is easily defined; being always clearly separable from the upper beds of the Silurian System, which decompose into a grey surface. In Shropshire and the contiguous parts of Herefordshire, this lower member of the Old Red System rarely occupies high ground, (except in the instance of the outlier of Clun Forest, hereafter to be described) and being for the most part recumbent on the talus of the upper Silurian rocks, where the latter sink down into valleys, it is generally much obscured by alluvial detritus. In the gorge of the Teme, however, between Ludlow and Downton Castle, it is well laid open, particularly at a spot called the Tin Mill. Flaglike, micaceous, dark red sandstone 'Bur Stones' rise there at an angle of about fifteen degrees from beneath the red argillaceous marls of Oakley Park, and pass down into a lightish-coloured grey, yellowish, and greenish grey freestone, of which Downton Castle is built, which will presently be described as constituting the upper stratum of the Silurian System. Similar relations are visible at Ludlow, and at Richard's Castle to the south of Ludlow.
>
> In this district, however, these lower red and yellowish beds, or 'bur stones', are seldom so fissile as the 'tile stones' described in South Wales ...

> Tilestone Group, east side of Herefordshire. As the Old Red Sandstone lies in a vast trough bounded by the Silurian System both on its eastern and western flanks, we ought to find its lower member, or tilestones, forming the western fringe of the Malvern Hills. Owing, however, to high inclination, the accumulation of detritus, and other results of disturbance, these beds are rarely well displayed for any distance along the eastern frontier of the Herefordshire basin. They are, however, clearly laid bare in a natural transverse section at Brockhill Knell between Mathon and Ledbury, where thin bands of yellowish green, micaceous flagstone, one and two inches thick, are subordinate to red, green and purple marls, the whole dipping away to the west and overlying the grey Ludlow rocks at an angle of forty-five degrees. Hard and thin flaggy rocks belonging to this group are also seen at the north-eastern suburb of Ledbury, dipping fifty-five degrees west-north-west, but the flanks of the ledges of older rocks near that town are encumbered with so much stiff red clay and detritus that the exact junction beds can rarely be distinguished. The same causes of obscuration, apply to the line of junction between the Old Red Sandstone and the Sulurian rocks of the May Hill range. In some valleys of elevation, however, the upper surfaces of the grey-coloured Silurian Rocks, which are thrown up in their interior, exhibit on their external faces clear examples of passage into the bottom beds of the Old Red Sandstone. This is well seen on the eastern slopes of the Clytha Hills, two or three miles east of Ragland, and will be further alluded to in the sequel. (Murchison 1839, 181–3)

Earp and Hains refer to 'the Tilestones-Downton Castle Sandstone horizon which provided some of the best roofing tiles' (Earp and Hains 1971, 103).

The Tilestones have been delved extensively throughout their exposure in Wales and eastwards to the vicinity of Downton Castle. Historically, sources from Clyro Hill to Huntington Hill, Gladestry and Kington have been important.

Speaking of Shropshire, Joseph Plymley noted that 'Very good stone slates for covering roofs are met with in the parish of Bettus on the south west confines of the county' (Plymley 1803, 67–8). Bettus is taken to be Bettws y Crwyn but may refer specifically to delves further east near New Invention (SO 278752) which would probably have been the source of roofing for Clun and Knighton (Figs 65 and 77) (Andrew Jenkinson, personal communication).

Brownstone Group

> The Brownstones form the highest division of the Lower Old Red Sandstone and consist largely of dingy purple-grey micaceous sandstone with some bands of red and green marl ... Some of the sandstones are massive, others flaggy, and both types have been extensively worked for building, paving and roofing. (Welch and Trotter 1961, 33)

Within the region the Brownstone Group is mainly exposed in the south surrounding the Forest of Dean but no delves specifically worked for stone slates are known.

Figure 77. Sandstone slate from Mainstone Church near Bettws y Crwyn, Shropshire. Munsell 2.5Y (© Terry Hughes).

Carboniferous

Within this region, the rocks of Carboniferous age extend around and to the north of Bristol, in the Forest of Dean, around the Wyre Forest at Kidderminster and northwards through Telford, Shrewsbury and Oswestry to the coalfields of Wrexham and Point of Ayr, although only the first two locations are known as stone slate sources. The Pennant sandstone has been the most significant source of roofing from this geological period in this region and the source of the same stone to the north of the South Wales coalfield may be significant for the future (Fig 78).

It should be noted that the name 'Pennant' has been used in a variety of ways in different parts of this region and a system developed to designate different formations. Explanations are given in Welch and Trotter (1961, 92–3) and Green (1992, 51).

Bristol Region

Thomas Rudge described the production of stone slates around Iron Acton north-east of Bristol.

> The colour of these [Cotswolds slates] are yellow or grey; but another sort of red grit is dug at Iron Acton and some adjoining places; as these, however, separate into thick lamina, and of course require strong timbers for their support, they are less eligible than the former. (Rudge 1813, 24)

Rudge's red grit is the red Pennant sandstone (Figs 79 and 80) that was extensively worked in many small delves around Iron Acton. In some locations the fissile stone is

Figure 78. A Carboniferous Pennant sandstone slate from the South Wales coalfield (© Terry Hughes).

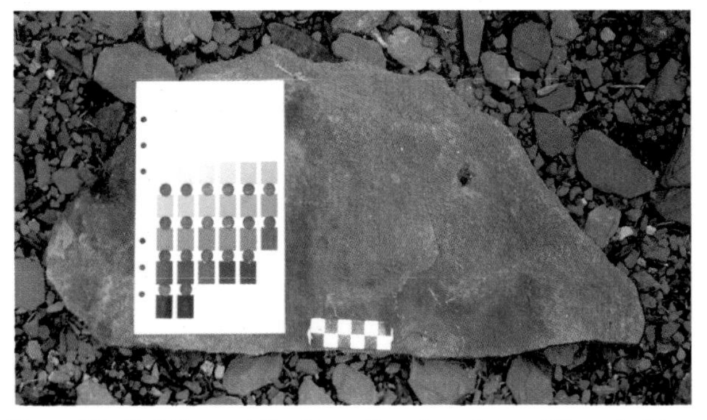

Figure 79. Red Pennant Sandstone from Iron Acton, Gloucestershire (© Terry Hughes).

Figure 80. Red Pennant Sandstone formerly quarried near Iron Acton, Gloucestershire (© Terry Hughes).

Figure 81. Pennant sandstone from Gwrhyd Specialist Stone Quarry in South Wales (© Terry Hughes).

close to the surface but at others it is found below several metres of clay. In the latter situations the stone was sometimes worked into the sides of valleys. It has not been possible to identify specific stone slate delves (often these were small and temporary) but there were more delves 'at Frampton Cotterall, Mangotsfield, Stapleton, Winterbourne' (Davey 1976, 23) and Temple Cloud. The Old Series Ordnance Survey map of the Bristol region (facsimile David & Charles edition, which includes revisions up to the 1840s) shows Tompkin's Quarry in the Carboniferous to the east of Iron Acton and Tillet's to the west but by the 1889 edition neither were recorded.

Nearer to Bristol, there were many workings in the Pennant sandstone along the river Avon from 'Netham to Hanham and in the Frome valley above Stapleton and at Mangotsfield and Fishponds. In former days Pennant sandstone was used on a very large scale for shaped building stone, kerbs and paving' (Kellaway and Welch 1993, 157) and would have also produced stone slates. Excavations of an old stone cutting workshop, formerly a mill, near Fishponds, have produced Pennant stone slates (John Penny, personal communication). There appear to be few Pennant roofs remaining in the region and currently no local source of supply but Gwrhyd Quarry in South Wales is now producing roofing (Fig 81), albeit not red. One roof, St Mathias' Chapel at the University of the West of England, was recently partially re-roofed with Carmyllie stone slates

(Devonian, Old Red Sandstone) from Tayside in Scotland and partially with a South Wales Pennant.

It has been reported that the Park Pennant Sandstone in Conygar quarry near Clevedon Court has fissile rock in the upper layers which could be quarried for roofing if required (Paul Chapman, personal communication).

Forest of Dean

> the Romans quarried and used local stone for their buildings both at Lydney Park and at the Chesters Villa in Woolaston, where were found stone roofing tiles and sandstone flags. (Hart 1971, 296)

Geologically, the Forest consists of a basin, elongated north-south, the central part of which contains Silesian Coal Measures. A thin rim of Dinantian limestone and dolomite surrounds the Silesian rocks, and the whole Carboniferous outcrop takes the form of an outlier resting on Devonian strata. Both Lower Devonian (the Lower Old Red Sandstone) and Upper Devonian (Upper Old Red Sandstone) strata are present, including the important sandstone roofing sources of the Dittonian and Breconian. The Upper Old Red Sandstone horizon passes transitionally into the basal Carboniferous forming an outer concentric rim around the carboniferous outcrop. The two Devonian formations were also exploited for roofing throughout the southern Marches but no specific roofing delves have been found within the Forest, although many include fissile rock. Within the Carboniferous, the Pennant sandstone appears to have been the only source of roofing (Fig 82).

> Pennant sandstone from the Coal Measures, varied in texture and colour, usually in shades of blue and grey. The best Blue Stone beds are found below the Grey, some 100-200ft below the surface according to the distance from the outcrop of Deans geological basin-outcrop. There are usually four layers of Blue Stone varying from 2ft 6in to 5ft 6in in thickness; it is much superior to the Grey, in colour and texture, and commands a higher price as monumental and better class building stone. (Hart 1971, 306)

The quarrying and mining industry within the Forest has a very long history and was very large in the past, exploiting sources of coal, iron, limestone and sandstone. Many of the sandstone sites would have produced stone slates from time to time although there are few records which refer to them specifically. The Forest inhabitant's right to work stone, as distinct from coal and iron which is a privilege of Free Miners, was confirmed and formalized by the Dean Forest (Mines) Act 1838. This, and earlier investigations and disputes, describe the products of the industry, including passing references to roofing stone.

> On 5th March 1661 Commissioners including Lord Herbert, constable and warden, appointed to enquire into the state and condition of Dean, were presented by the inhabitants with a 'Memorial' claiming 'as enjoyed for divers hundreds of years' *inter alia*, liberty to dig and get limestone, tilestone and other stones necessary to be employed in and upon their ancient messuages, lands and tenements, and also to make and get millstones and grindstones. (Hart 1971, 299) [33]

Figure 82. Pennant sandstone on the roof of St Briavels Castle, Forest of Dean (© Terry Hughes).

The Commissioners reported to the Exchequer 12 April 1662 (BM Harl. MSS 6839, fol 335):

> There are several quarries of stone and grindstone fit for the Navy, paving, tyle, slate, excellent for building and some Millstone, but by reason that never as yet any yearly advantage accrued by it, we cannot set upon it any positive value. (Hart 1971, 299)

Today the stone industry of the Forest is small and, although there are two planning consents, both in the Pennant, which specifically mention stone slate production (at The Old Flour Mill, Bream and Meezy Hurst Quarry, Staple Edge Wood), neither are currently producing roofing.

Permian

D C Davis, speaking of roofing stones, recorded that 'the thin calcareous flags that lie between the Bunter and Keuper sandstones in Shropshire, at Grinshill, were formerly used for the same purpose' (Davis 1912, 77). This appears to have been a minor roofing stone although important locally. The specific source, the Grinshill Flagstone overlying the White Grinshill Sandstone, has been described by Thompson (1993 and 1995). The fissile stone can be seen high up in the wall of Deep Quarry. The bed thickness ranges from 450 mm down, the thinnest being used for roofing. They split along the clay layers but are not a durable stone (D Thompson, personal communication).

Jurassic

It is possible that stone from the base of the Lias was used for roofing at Ashleworth and around Tewkesbury. Suitable stone has been found in field walls in the area although at present no examples are known on roofs. There are records of old quarries at Sarn Hill, Hill Croombe and Heath Hill near Tewkesbury (Richardson 1904, 36–41) which are all within a mile or two of the

Figure 83. Rarely, stone slates can be made from igneous rock. This is a dolerite from Corndon Hill in Shropshire (© Terry Hughes).

River Severn which would have provided an easy route to Ashleworth. Murchison (1839, 21) recorded the colloquial names of the six beds quarried in the area: Top, Black, Tile, Poacher, Peaver and Bottom. Unfortunately, as is so often the case, it is not clear whether Tile implies a roofing or flooring application. He also noted 'In descending order we first perceive about 12 thin courses of dark-coloured calcareous flagstones, which are extracted for roads, paving, building, and burning to lime. These courses vary in thickness from 1 to 3 inches, and are separated from each other by stiff marl' (Murchison 1839, 20). Other exposures of the rock indicate that the stone could have been worked along both sides of the river from north of Tewsbury southwards to at least Ashleworth (David Jefferson and Graham Lott, personal communication).

Howe reported a source of stone slates from the Lower Lias at Burley Dam east of Whitchurch in Cheshire (Howe 1910, 317).[34] He was probably quoting Murchison who stated 'In the vicinity of Burley Dam, some of the [Lower Lias] beds are so hard as to have induced Lord Combermere to quarry them for slating purposes' (Murchison 1839, 23).

Whatever the intentions of Lord Combermere there is no Lias roofing at Combermere Abbey now (Gerald Emerton, personal communication). So it has to be assumed for the present that the initiative either came to nothing or the slates eventually reached the end of their life and were replaced.

Igneous

> Very good stone slates for covering roofs are met with in the parish of Bettus on the south west confines of the county and there is very good flagstone in Corndon Hill west of Bishop's Castle. (Plymley 1803, 67–8)

North reviewed the use of volcanic rock from Corndon Hill, north of Bishops Castle.

> Some flaggy rocks of rather unusual character are associated with the massive igneous (intrusive) rocks of Corndon Hill, near the Shropshire border of Montgomeryshire. They were first described by Murchison in his great work on the Silurian System. He wrote (in 1839) 'on the west and south-western face of the summit of the Corndon, the greenstone [we should now say dolerite] graduates upwards into a thinly laminated rock, dipping to the west at an angle of about thirty-five degrees, which has been worked for flagstones. These flaggy beds ... are of a dull grey colour, have a talcose and saponaceous feel, and a rough mechanical cross fracture'. He suggested that the beds represented material thrown out during submarine eruptions at an early stage in the volcanic episode which affected the area during what we should now call Ordovician times.
>
> The great antiquity of the Corndon quarry and of the use of its stone is indicated by the fact, noted by Peate in The Welsh House, that the mid 15C poet, Guto'r Glyn, described the house of a vicar of Llandrinio, in Montgomeryshire, as having Cornatun (Corndon) stone on its roof. (North 1946, 54)

Corndon Quarry can be approached from the south through Old Chirbury. There are shallow surface workings over a wide area which appear to have followed the rock uphill without increase in overburden. There are lots of pieces of rock lying around up to 20 inches by 18 inches (500 x 450 mm), about one inch thick (25 mm), and they are all sound and un-delaminated (Fig 83). None of the roofs in the immediate area are stone but some of the older roofs in the region may still contain some Corndon stone.

Market

The market for the whole region is substantial, especially as the roofs of so many churches are currently reaching the end of their lives. In respect of the individual stone types, the markets are modest and in many instances may not support continuous production. How the market develops will depend on the strategy adopted by the building conservation bodies. Three steps are essential: identify the specific stone slate types, estimate the market size for each, and establish an appropriate level and form of manufacture integrated with a programme of grant support and conservation control.

Production

Currently there are two delves producing stone slates in the region and in spite of great efforts to establish other sources of production in the area south of Hay-on-Wye there has until recently been no progress. Now, a community initiative in the Golden Valley shows every prospect of being successful. The intention is to re-establish production in several small old delves, possibly each supplying stone to a central processing point. Initially the stone slates will be used to re-roof Dore Abbey.

In the Forest of Dean there are two planning consents, both in the Pennant, which specifically mention stone slate production, but neither are currently producing roofing. Pennant sandstone is now being delved for roofing near Pontardawe, in South Wales.

The big success in the region has been the delving of Harnage stone slates specifically for Pitchford Church and Hall (see Wood and Hughes, this volume). This is an example of how important and distinctive roof types can be conserved if there is sufficient commitment from the relevant conservation and control bodies. It represents an initiative on the part of English Heritage which goes to the heart of the problem of conserving stone buildings: that without access to the original stone and consent for delving, the objective cannot be achieved. The project turned out to be very difficult, although most of the difficulties were associated with people and systems rather than technical issues, and required a sustained effort on the part of those most closely involved. Much was learned which will benefit similar initiatives in the future.

Delves producing slate in the Welsh Marches and Bristol region are listed in Annex A.

Other sources

Despite the best efforts of conservation bodies, architects and others involved in building conservation, authentic Old Red Sandstone and Ordovician (Harnage excepted) roofs in the region are disappearing at an alarming rate. For many years only reclaimed stone slates were available for authentic re-roofing. The shortfall is being made up with Carboniferous sandstone slate from sources outside the region, from Wales and northern England. In contrast, Pennant and Harnage roofs can now be renewed with the correct slates. For south-west Hereford at least, the situation has improved recently.

Summary

The stone slate roofscape of this region is geologically the most diverse and interesting in England. Unfortunately, because some of the slate types are not being manufactured at present, there is a real risk that most of these roofs which contribute so much to the regional distinctiveness will eventually be lost. Urgent action is required to develop a regional strategy, including Wales, for tackling the problems of production.

DERBYSHIRE AND THE PEAK PARK

> Slates or Tilestones: these, in the district where the lamellar stones abound, are mostly used instead of Tiles; or blue slates for the Houses and Buildings. At Sheffield these white and grey Slates are exclusively used, and give the Town a novel appearance to a stranger approaching it by the Mansfield road. Most of the grey slate of this district abounds with Mica in minute plates, forming layers at the joints where the stone most readily parts. In numerous instances these joints are remarkably plane and smooth, but in others the surface of the Slates are waved and curled in a very regular and curious manner: these waved Slates, altho' they are seldom so light or look so well on a house as the plain ones, are nevertheless found to last, and answer the best in many situations. (Farey 1811, 428)

The Carboniferous sandstones have provided roofing material throughout the region and into neighbouring counties, and a substantial market for a variety of local types still exists which, until recently, had little prospect of being satisfied. There was also a much smaller use of rock from the Magnesian Limestone sequence around Whitwell in east Derbyshire. Farey (1811, 428–31) listed the sources of stones in Derbyshire. Stanley has correlated most of these to modern locations (Stanley 1993).

Geology

Within this region, stone slates are almost entirely derived from Carboniferous rocks of Namurian and Westphalian age. The only known exception is rock from the Magnesian Limestone in east Derbyshire and Nottinghamshire (Table 18).

For historical reasons, the individual geological units which have produced stone slates have different names across the region and these are set out in Table 19 in relation to sub-areas within the region. These units are not all equally important as roofing sources. The sub-areas in Figure 84 are those used to analyse the combined geological and geographical distribution in the study of roofing stones in the South Pennines (Hughes *et al* 1995). They are based upon a combination of similar geological features, clusters of former workings, specific types of stone slates and former or potential markets. They have no formal status and their boundaries are necessarily broadly drawn; their limits frequently follow natural watersheds and are usually in sparsely populated areas lacking potential stone slates.

Geologically, the region has the form of an inverted U with the Carboniferous limestone forming the centre. The beds of Namurian age, originally termed the Millstone Grit Series in this region, surround the limestone and pass northwards into Yorkshire, Lancashire, Cumbria, County Durham and Northumberland. To the east and west, they are flanked by the Coal Measures of Westphalian age and, further to the east, by the Magnesian Limestone of Permian age. The locations and the geological units which have been worked in the Carboniferous are shown in Table 19.

The large number of rocks which have been used for roofing implies that there are a similar number of stone slate types. However, many are similar in geology and appearance and the study carried out into the grey slates in this region (Hughes 1996) concluded, on an amalgam of geological and visual criteria, that seven types would reasonably cover the full range (see Figs 11–18). They are named for a type location but, because they are a product of geology, they may have a wide distribution in the region (Table 20). The Cracken Edge type is overwhelmingly the most common within the region.

The geology and the historical sources of stone slates in the region which were researched as part of the regional study (Hughes 1996), were reviewed by Ian Thomas in Hughes *et al* 1995, B1–B37. It is reproduced here in an edited form. The location of the sub-areas are shown in Figure 84.

Carboniferous

Dane Valley and Staffordshire Moors

This area is characterized by ridge upon ridge of mainly north-south trending sandstones, separated by mudstone

Table 18. Namurian, Westphalian and Permian sources of stone-slates within Derbyshire and the Peak Park.

Age				Rock units
Permian	Upper	Cadeby Formation Marl Slate		Lower Magnesian Limestone
	Lower	Basal Permian sands and breccias		
Upper Carboniferous (Silesian)	Series	Stage		
	Stephanian	C B A	Probably absent	
	Westphalian	D	Upper Coal Measures	
		C	Upper and part of the Middle Coal Measures	
		B	Middle Coal Measures	
		A	Lower Coal Measures	Wingfield Flags, Silkstone Rock, Penistone Flags, Grenoside Sandstone, Greenmore Rock, Brincliffe, Edge Rock, Loxley Edge Rock, Upper Band Rock, Milnrow Sandstone, Crawshaw Sandstone, Woodhead Hill Rock, Sandstones below the Red Ash Coal, Sandstones above the Yard Coal
		G_1	Yeadonian	Rough Rock Rough Rock Flags
	Namurian – Millstone Grit 'Series'	R_2	Marsdenian	Huddersfield White Rock, Beacon Hill Flags, Pule Hill Grit, Heydon Rock, Rivilin Grit, Chatsworth Grit, Readycon Dean Series, Redmires Flags, Roaches Grit, Corbar Grit, Ashover Grit, Five Clouds Sandstone, Brown Edge Flags, Rushtonhall Grit, Walker Barn Grit.
		R_1	Kinderscoutian	Kinderscout Grit, Shale Grit, Edale Shales
		H_2	Alportian	Edale Shales
		H_1	Chokerian	Edale Shales
		E_2	Arnsbergian	Edale Shales
		E_1	Pendleian	Edale Shales

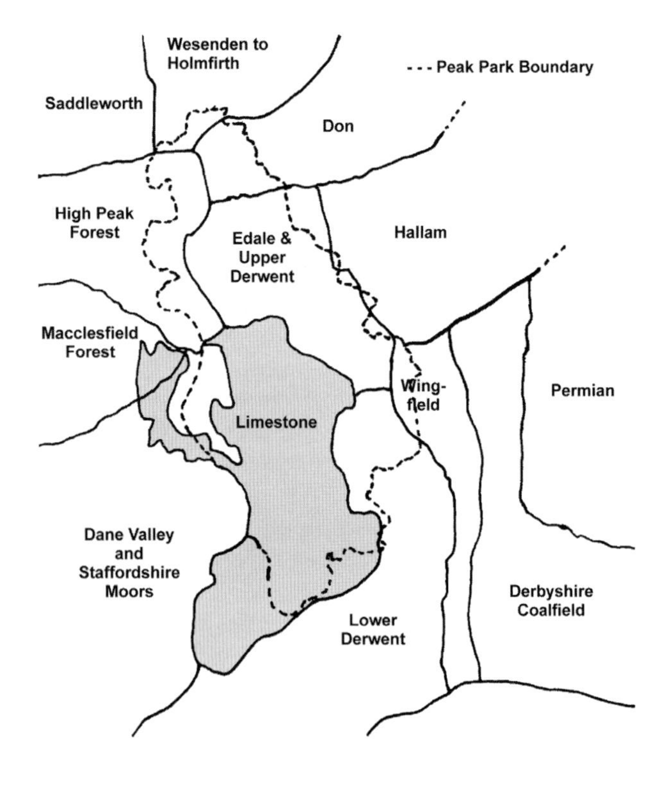

Figure 84. The sub-areas used to analyse the combined geological and geographical distribution of roofing stone in the South Pennines study (Hughes et al *1995).*

valleys. Some of the moors, notably Axe Edge and Morridge, form major tracts of high peat-covered hills. In contrast, the Roaches and Hen Cloud area comprises a series of dramatic rocky ridges. The western part of the area is more subdued.

By far the most important source of stone slates is the part where the twin operations of Danebower and Reeve Edge (Fig 85) worked the Rough Rock on either side of the Upper Dane Valley. Despite the relatively remote location, these were among the three or four most extensive operations located in the study and appear to have been very largely devoted to stone slate production. Two or three smaller workings were also operative in the Rough Rock nearby and the local name Bakestone Edge is also pertinent here.

The other traditional sources located were Daisy Knowle Mine, Longnor (Longnor Sandstone), the only totally underground operation, and small sections of Five Clouds Quarry overlooked by the Roaches (Fig 86). Both these sources were worked in fairly arduous circumstances.

Macclesfield Forest

This area is typified by a concentration of narrow, relatively steeply-dipping north-south trending sandstone ridges, making up the tightly-folded Goyt syncline in the east and the Todd Brook anticline further west. Beyond,

Table 19. Locations of stone slates in Derbyshire and the Peak District National Park.

Age	Rock unit	County	Sub-areas and locations
			Dane Valley and Staffordshire Moors
Namurian	Rough Rock	Cheshire	Danebower and Reeve Edge
Namurian	Longnor Sst	Staffs	Longnor
	Five Clouds Sst	Staffs	Roaches Five Clouds
			Macclesfield Forest
Westphalian	Milnrow Sst	Cheshire	Kerridge, Bakestonedale Moor
Namurian	Rough Rock	Cheshire	Billinge Hill
Namurian	Rough Rock	Derbys	Goytsclough
Namurian	Chatsworth Grit	Cheshire	Teggs Nose
Namurian	Chatsworth Grit	Derbys	Wingather Rocks
			High Peak Forest
Namurian	Kinderscout Grit	Derbys	Chinley to Chinley Head, Chunal, Whitfield, Glossop Low
Namurian	Rough Rock	Derbys	Cracken Edge, Chinley Churn, Rowarth
Namurian	Chatsworth Grit	Derbys	Buxworth to Eccles Pike
Namurian	Chatsworth Grit & Kinderscout Grit	Derbys	Hayfield-Birch Valley
Namurian	Kinderscout Grit	Derbys	Longdendale Valley: Tintwistle
			Saddleworth
Namurian	Kinderscout Grit	Lancs	Slate Pit Moor, Carrsbrook, Mossley
Namurian	Kinderscout Grit	Yorks	Diggle: Ravenstone Rocks Quarry
Namurian	Kinderscout Grit		Pennine Moors: Slate Pit Moss, Far and Near Broadslate
			Wissenden - Holmfirth
Namurian	Pule Hill Grit	Yorks	Meltham: Scouthill Quarry
Namurian	Huddersfield White Rock	Yorks	Meltham Moor: Royd Edge and Isle of Skye quarries, Holme Valley
			Don
Namurian	Rough Rock Flags	Yorks	Hade Edge to Winscar reservoir, Magnum Bonum & Snailsden Moss, Winscar reservoir: Harden Edge and Tyas Quarry
Westphalian	Grenoside Rock	Yorks	Hartcliffe
			Hallam
Namurian	Heydon Rock	Yorks	Thornseat Delf
Namurian	Huddersfield White Rock	Yorks	Loadfield Quarry
Namurian	Rough Rock	Yorks	Oughtibridge, Spout House Hill, Kirk Edge
Namurian	Rough Rock	Yorks	Kirk Edge (Slates?), Ringinglow: Fulwood Booth and Brown Edge, Whirlow Bridge (Sheffield)
Namurian	Rough Rock	Derbys	White Edge Moor near Nether Padley
			Edale and Upper Derwent
Namurian	Shale Grit	Derbys	Abney: Shatton Moor
Namurian	Kinderscout Grit	Derbys	Wet Withens, Bamford
			Lower Derwent
	Ashover Grit	Derbys	Bakewell Edge, White Tor near Starkholmes
			Wingfield Flags
Westphalian	Greenmoor Rock	Yorks	Totley-Cordwell
	Wingfield Flags	Derbys	Bolehill Wingerworth, Freebirch, Walton (Slatepit Dale) to Clay Cross
			Derbyshire Coalfield
Westphalian	Ssts of the Top Hard & High Hazels coals	Derbys	Temple Normanton, Sutton Scarsdale,
Westphalian	Deep Soft coal	Derbys	Tibshelf
Westphalian		Yorks	Teversal
Westphalian	Silkstone Rock – Wingfield Flags	Derbys	Cartledge
	Deep Hard coal	Derbys	Tupton to Newbold

Figure 85. Sandstone slate from Reeve Edge. Munsell 2.5YR (© Oxford Brookes University).

Figure 86. Derbyshire and the Peak Park: a roof at the Roaches in the Dane Valley and Staffordshire Moorlands sub-area (© Terry Hughes).

Figure 87. Derbyshire and the Peak Park: a roof at Flash in the Dane Valley and Staffordshire Moorlands sub-area (© Terry Hughes).

in the north west (east of Pott Shrigley) is a series of east-west aligned sandstones. The folding results in repeated outcrops of the same horizons on each side of the folds, and all the key sandstones found in the study area are represented here. Almost all have been quarried (Fig 87).

A number of workings, some on a large scale, were operated here. Immediately outside the Peak Park, an almost continuous series of quarries in the Milnrow sandstone can be followed for 2 km (1.24 miles) along the western side of Kerridge Hill, some of which are still active and one, Macstone quarry, which has recently attempted production of stone slates (see Fig 12).

The Milnrow Sandstone was also quarried until very recently at Bakestonedale Moor; older workings extend eastwards. Much of the rock here has a superficially flaggy appearance, but the survey failed to identify any roofing material. Billinge Hill Quarries and the historically important Goytsclough Quarry relied upon the Rough Rock. Although the latter was referred to by Farey as a stone slate source, insufficient suitable material was found in the survey and the overburden proved to be too great to merit working, notwithstanding possible objections on account of its sensitive location. Teggs Nose Quarry (also now a country park) exposes an untypical turbidite sequence (alternating thin blocky sandstones and mudstones) in the Chatsworth Grit. Although listed by Farey as a source, the proportion of fissile rock in the exposed sections suggests that it was not a very important source.

High Peak Forest

The great bulk of the classic peat-clad Dark Peak comprising Kinder Scout, Bleaklow and so on, defines the eastern boundary. It is formed by great alternating sequences of thin sandstone and shale, making up the Shale Grit (a term coined by Farey) and the normally coarsely-grained massively bedded Kinderscout Grit. To the west, the beds of the thin outcrops of Chatsworth Grit, Rough Rock and Lower Westphalian Sandstones account for numerous fault-broken north-south trending ridges.

Table 20. Derbyshire and Peak Park stone slate types and their occurrence.

Type	Features	Sub-areas	Figure
Yorkstone	flat, featureless, without substantial stepped bedding, fine- to medium-grained, buff to dark brown	Wessenden – Holmfirth, Don, Hallam, Derbyshire Coalfield	11
Kerridge	flat, featureless, without stepped bedding, fine-grained, grey mica surface	Macclesfield Common	12
Cracken Edge	textured, with or without stepped bedding, fine- to coarse-grained, white and buff to dark brown	High Peak Forest, Macclesfield Common, Edale and Upper Derwent	13
Teggs Nose	textured, with or without stepped bedding, fine- to medium-grained, pink	High Peak Forest, Macclesfield Common	14
Freebirch	strongly textured or ripple etc marked, fine- to medium-grained, buff to dark brown and olive to grey	Lower Derwent, Wingfield, Derbyshire Coalfield	15
Wirksworth	strongly textured, fine- to medium-grained, pink to red	Lower Derwent	16
Whitwell	strongly textured or ripple etc marked, fine-grained, grey or pink, magnesian limestone	Permian	17 and 18

Figure 88. Derbyshire and the Peak Park: Tintwistle near Glossop in the High Peak Forest sub-area (© Terry Hughes).

Occasionally, where the Kinderscout Grit has been separated by bands of intervening mudstone into a series of subsidiary leaves, its characteristics change to a more laminated, finer-grained style, as in the former stone slate workings along the A624 from Chinley to Chinley Head, Chunal and Whitfield, (now part of Glossop), all in the lower leaf. However, the most notable and one of the most isolated sites, that at Glossop Low, utilized the Middle Leaf. These quarries and large scale tips extend for about a kilometre (0.62 miles) across open moorland on the Norfolk Estate and had their heyday in the period 1800–50. These constitute some of the largest undertakings seen in the study. Cracken Edge on Chinley Churn is the other large and long-standing stone slate production site. Here, the Rough Rock has been worked, occasionally by underground mining, over a distance of nearly 2 km (1.24 miles) and involved a more mechanized approach than elsewhere, finally closing in about 1920. The Rough Rock is also worked further north, particularly around Rowarth but on a smaller scale and also for building stone. Between these two, the Chatsworth Grit, although rarely acceptable for roofing elsewhere, appears to be of particularly good quality between Buxworth and Eccles Pike. In this area, the higher, near surface sections are of 'well cleaved' thin false-bedded sandstones. In contrast, other quarries in the Chatsworth Grit and Kinderscout Grit in the Hayfield-Birch Valley area, reveal the massive coarse-grained characteristics more typical of these sandstones. Even the Rough Rock is not particularly flaggy here. In the main part of the Longdendale Valley, at least three quarries in the main bed of the Kinderscout Grit produced good stone slate material at Tintwistle (Fig 88).

Saddleworth

In this sub-area, the north-south trending Tame Valley, with its ribbon of mill towns, Mossley, Upper Mill and Diggle, divides the peat-covered Kinderscout Grit moors

Figure 89. Throughout the Pennines it is common for a single delph to have produced every part of a building. Here the walls, door jambs, lintels, cills, flooring and roof have all come from a delph in the Holme Valley (© Terry Hughes).

of the Pennines from the Lancashire Coalfield to the West. Shale Grit with some Boulder Clay cover forms the floors of the valleys, with the Kinderscout Grit generally outcropping on the higher eastern slopes. Much of the latter is typical coarse, even pebbly, massively bedded material and is sufficiently cohesive to be worked for aggregates at Buckton Moor Quarry. However, the main Grit is separated by extensive mudstone bands, in places into at least four separate sandstone units.

In this area, a former quarrying operation worked the middle leaf of the Lower Kinderscout Grit, the base of which is flaggy. The site at Slate Pit Moor, Carrbrook, Mossley at the top of a steep valley-side appears to have been reached by means of an incline. There is further evidence for stone slate working in extremely remote parts of the Pennine Moors, notably at Slate Pit Moss and at Far and Near Broadslate. Higher sections of the Kinderscout Grit, and the Readycon Dean Series underlie these areas.

Wessenden to Holmfirth

This area is complex geologically and topographically. It includes the main Pennine ridge but this is fragmented by important subsidiary valleys, some fault-controlled but having no consistent pattern. Kinderscout Grit accounts for much of the higher land. Younger sandstone units, notably the Readycon Dean Series, Huddersfield White Rock (Middle Grits/Marsdenian) and the Rough Rock also underlie extensive high land to the east and the more populated, rather lower plateaux and valleys from Meltham to Holmfirth. These and other sandstones have almost all been worked, often extensively for building stones. The Pule Hill Grit produced flags and possibly stone slates at Scouthill Quarry, Meltham, as did the Huddersfield White Rock on Meltham Moor (Royd Edge and Isle of Skye quarries) and on either side of the Holme Valley (Fig 89).

However, despite the extensive use of stone slates in many of the older buildings, notably at Holme, no major sources clearly emerged. It is possible that most of the supplies were drawn from the many widely quarried sandstones in the more urbanized lower land, just outside the study area, ie towards Huddersfield and Brighouse to the north and at Harden Clough in the Don sub-area. One notable feature is the coarse and massive nature of the Rough Rock in the large outlier, west of Meltham,

Figure 90. Sandstone from Harden Clough showing bioturbation. The small, dark lumps are casts of sand deposited by filter feeding burrowing worms. Munsell 2.5Y (© Terry Hughes).

Figure 91. In the Hallam sub-area of Derbyshire and the Peak Park the Coal Measures were extensively worked for slates. This example is at Mickley near Sheffield. Open-cast coal workings in this area have removed all trace of many of the old quarries (© Terry Hughes).

in contrast to the abundance of fine-grained fissile material in the sandstone to the south east of this sub-area.

Don

This sub-area comprises the Don and Little Don Valleys, from Stocksbridge and Penistone to their headwaters. Together, they divide the Yorkshire Coalfield containing the Greenmoor Rock and Grenoside Sandstone (equivalent to the Wingfield Flags) to the north east, from the Rough Rock and Huddersfield White Rock rising to the Kinderscout Grit, south of the Woodhead Pass. Locally, in the west, the base of the Rough Rock has been mapped as a separate sub-unit, the Rough Rock Flags. Between Hade Edge and Winscar Reservoir, these Flags have been exploited in conjunction with the intervening Rough Rock Coal over an area of 1.5 km by 0.5 km (0.93 by 0.3 miles) either side of the Harden Clough (Fig 90). Operations continued here at the so-called Magnum Bonum Quarries into the 1920s and with Snailsden Moss to the south, produced stone slates, kerbs, flags and building stone. Nearby, on the east side of Winscar Reservoir, two small workings below Harden Edge and Tyas Quarry worked the same sub-unit for stone slates.

From Flouch Inn to Midhopestones, despite the preponderance of flaggy stone walls and stone slate roofs in the area, and the continuing presence of the Rough Rock, there is no evident local source in these rocks. Indeed, they are virtually devoid of quarries. The Rough Rock Flags do die out in this direction. Possibly workings were shallow and widely scattered across the outcrop. Alternatively, Hartcliffe (just outside the area but listed by Farey), in the Grenoside Rock, may have been the supplier or Magnum Bonum to the north may have dominated the local market.

Hallam

The western half of this sub-area comprises high peat moorland largely on Kinderscout Grit. To the east, the sandstone of the Middle Grits (Marsdenian), Rough Rock and the lowermost Westphalian sandstones outcrop in fairly regular fashion, interrupted by faulting in the east. Attractive valleys with strings of reservoirs are a key feature of this part of the sub-area. The broad spread of the Heyden Rock and the Chatsworth Grit (locally called the Rivelin Grit) both supported isolated stone slate workings. Those in the lower part of the Heydon Rock at Thornseat Delf are on a large scale, covering about eight to nine hectares. The Huddersfield White Rock is still extracted at Loadfield Quarry and appears to have some potential for stone slates.

By contrast, south of Ewden Beck, the Rough Rock has a rather more fragmented outcrop but has been very widely quarried, for example from Oughtibridge to Spout House Hill, then south to Kirk Edge. Over a distance of 6 km (3.72 miles) there is an almost unbroken string of quarries. Those at Spout House Hill alone cover an area of 0.75 by 0.5 km (0.46 by 0.3 miles). They mainly produced stone slates and bakestones and appear to have been mined underground via shafts in the centre of the outcrop. Kirk Edge Quarries are rather enigmatic and may also have supplied stone slates as well as coarser material. The Rough Rock between Bradfield and Redmires Reservoirs does not appear to have been worked, but the same beds between the reservoirs and Ringinglow were very important sources. The operations at Fulwood Booth and Brown Edge are particularly significant and areas between these two places are worth further detailed investigation.

Further, but smaller, workings produced stone slates at Whirlow Bridge in Sheffield (Fig 91). However, other nearby outcrops of the Rough Rock, particularly to the north (west of Totley) and west (west of Redmires and on Ughill Moors) either tend to be coarse-grained or devoid of workings. A fairly small, detached outcrop at White Edge Moor, Nether Padley, was largely worked out for stone slates, but may offer some potential if overburden is not excessive. (This site is actually located in the Upper Derwent Valley but is more closely related to the Rough Rock of this sub-area.)

North of the Ewden Beck, the Rough Rock is remarkable for its lack of quarries in contrast to the extensive outcrop and the widespread use of stone slates

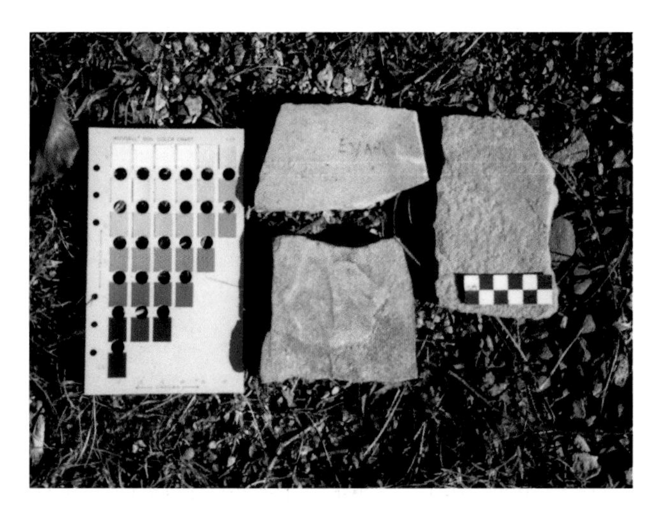

Figure 92. Samples quarried at Bretton near Eyam during production trials (© Terry Hughes).

Figure 93. The Lower Derwent sub-area of Derbyshire and the Peak Park has been a national source of building stone but there were few quarries with fissile stone. One of the few stone slates produced in the area is shown on this roof at Tansley near Matlock (© Terry Hughes).

in Bolsterstone village. The Crawshaw Sandstone and Loxley Edge Rock are typically medium- to coarse-grained and generally massive in character.

Edale and Upper Derwent

This sub-area takes in the Derwent Valley above Baslow and is dominated in the east by the classic Millstone Grit edges formed by the Kinderscout Grit and Chatsworth Grit to the north of Hathersage and the Chatsworth Grit and Crawshaw Sandstone to the south. West of the Derwent, the Kinderscout Grit and the Shale Grit contain the main sandstones and form virtually all the higher land. The valleys are floored by Edale Shales (which also fringes the Limestone) and the Mam Tor Beds. Apart from the detached upland centred on Abney, all the other gritstone moorland is part of the main Dark Peak plateau, dissected by the Derwent, Hope, Edale and Woodlands Valleys.

In the Abney area, Farey listed a number of sources and clearly there were probably at least a dozen shallow operations scattered across the flaggy leaves of the Shale Grit in the western part of the upland. However, they all appeared to have been opened up to serve very local needs, individual farms or for walling. In one instance, on Shatton Moor, five small workings exploited a single leaf of sandstone all linked to one access-way. If required, further detailed investigation might identify sufficient material from one of these sites, to serve local needs, particularly for the small slates typically used around Abney and Eyam (Fig 92). The overlying Kinderscout Grit was also exploited in similar but more limited fashion, the only workings of note being in the lower leaf of sandstone at Wet Withens. The lower detached leaf of the Kinderscout Grit was also worked at sites above Bamford.

Possibly the most remarkable feature of this sub-area is the dominant use of stone slates in the Vale of Edale since the 1650s but the lack of any clear evidence of their source. The extent of their use very strongly implies a local source. It may have been the case that the flaggy sections of the Shale Grit (there are numerous sandstone leaves) and over the Mam Tor Beds in the valley may have provided such an abundance of sources that no major suppliers were required. However, totally unlike Abney, there is no map and little field evidence to support this. The historical references to 'slate pits' in this area require further research to locate sites.

Lower Derwent

In addition to the main Derwent Valley south of Bakewell, this sub-area takes in the Ecclesbourne Valley (Fig 93). North of Cromford the Carboniferous Limestone constitutes the western boundary. Detached outliers of predominantly Ashover Grit make up Stanton Moor, Harthill and the fault-broken ridges of the area between the two main valleys. To the east of the Derwent, the Ashover and Chatsworth Grits typically form stepped scarps parallel to the main valley and around Ashover. Further east, the coarse Crawshaw Sandstone of the lower Westphalian outcrops, particularly on Holymoor and above Chatsworth. The intervening Rough Rock is thin or even absent in places but was a significant source of building stone at Coxbench.

Very few parts of the varied and extensive sandstone outcrops in the sub-area have been untouched by building stone operations, indeed, some of the sources could claim national significance, particularly in Victorian times and also in the last 20 years. In contrast the number of former stone slate sources is remarkably small. Farey referred to only six sites in this area, only two of which can be located with any certainty today. This suggests that the scale of such working that did take place was clearly very small and no other significant 'new' sites were found in this study. This view is supported by the infrequent evidence of the use of stone slates from remaining buildings; almost all of the latter are found close to the established sources in the area, namely Bakewell Edge and White Tor near Starkholmes. Field observation of most of the former building stone quarries also indicates that very few sec-

Figure 94. Around Heage at the southern end of the Wingfield sub-area of Derbyshire and the Peak Park a rough surfaced slate was used (© Terry Hughes).

Figure 95. The stone slates from Freebirch in the Wingfield sub-area of Derbyshire and the Peak Park have an unusually rough surface. The rippled surface shows that the sediment from which the rock was formed was laid down in a shallow water environment (© Terry Hughes).

tions could have produced stone slates despite a geological context apparently comparable to other parts of the area. The almost universal absence of fissile rock may have been one of the reasons which actually favoured the large-scale production of building stone here.

Wingfield Flags

This sub-area is the only one defined on the basis of its geology. Geographically it runs along the boundary between the classic gritstone moors (formed by the Namurian sandstone) east of the Derwent, and the main Derbyshire Coalfield. From Totley in the north to Kilburn in the south the outcrop is about 37 km (23 miles) long (Fig 94). The area is characterized by rolling farmland, relatively well wooded and by broken, generally north-south ridges. Throughout much of its length the Wingfield Flags almost invariably comprise a series of parallel sinuous sandstone outcrops; the intervening mudstones typically separate out between two and five sandstone bands. In the Totley-Cordwell Valley area, the beds are known as the Greenmoor Rock. The adjacent Westphalian sandstones often exhibit similar properties; they are almost invariably flaggy or at least thinly bedded, fine-grained sandstone or siltstone, micaceous in places and generally olive in colour. Fine ripple marking is a particular characteristic. The notable exception lies at Bolehill, Wingerworth where a more massive but still false-bedded buff building stone is quarried; flaggy material is confined to the very top of this section.

Elsewhere a dozen or so former stone slate workings are found in two clusters, around Freebirch (Fig 95) at the head of the Linacre Valley and between Slatepit Dale (Walton) (Fig 96) and Clay Cross. There are a number of possible sites for future working, mostly as extensions to former operations, particularly around Freebirch.

Although no workings could be found from site or documentary evidence at Totley, the geology of the area and the widespread use in roofs strongly implies that there were good local sources here.

Derbyshire Coalfield

The sub-area is defined on the west by the wooded sandstone ridges of the Wingfield Flags and on the east by

Figure 96. In the Wingfield sub-area of Derbyshire and the Peak Park this Millstone Grit roof at Old Brampton near Chesterfield is close to the boundary with the Coal Measures to the east (© Terry Hughes).

the Permian Limestone. Numerous cycles of coal seams, mudstones and sandstones underlie the area. The sandstones are especially variable, both in extent and nature, and are buff, yellow and occasionally pink or ochre in colour. Examples of stone slates are extremely rare, possibly in part due to mining subsidence followed by replacement with other materials. Many of the sandstones have been exploited on a limited scale mainly for (often poor) building stone but evidence on the ground to support documentary records has often been lost through opencast coal mining.

Most of the sites identified are concentrated in the area bounded by Temple Normanton, Sutton Scarsdale, Tibshelf and Teversal. Sandstones associated with the Top Hard and High Hazels Coal seams, although relatively thin, were most commonly worked. However this research was heavily dependent upon leads provided by historical and geological documentation which does not appear to be consistent across the Coalfield.

Elsewhere, in those fringe areas of the Coalfield where coal mining ceased many years ago, between the Silkstone Rock and the Greenmoor Rock/Wingfield Flags, stone slates (similar to the nearby Wingfield Flags) were extracted at Cartledge near Holmesfield and contributed to

Figure 97. In the Derbyshire Coal Measures sub-area of Derbyshire and the Peak Park most roofs are similar to that in Figure 91. The calcareous stone slates from the Magnesian Limestone are distinctively different in texture and the colour of the plants they support. This example is at Steetley near Chesterfield (© Terry Hughes).

a fine suite of roofs here. Further investigations are likely to identify a workable deposit. The Silkstone Rock itself has a broad outcrop from Dronfield Woodhouse to Eckington. Most of the former sandstone workings have been developed for housing. Small quarries in this sandstone at Peakly Hill may offer a source. Some of these operations produced stone slates. On a more limited scale, the sandstone overlying the Deep Hard Coal was worked between Tupton and Newbold, one site being specifically named Newbold Slate Pit.

Permian

Magnesian Limestone

The Magnesian Limestone extends to the east of the Carboniferous from Long Eaton to the coast at Sunderland but the only known use of it as roofing is around Whitwell in East Derbyshire, near Bolsover (see Fig 17) and at Gildingwells Quarry, north of Worksop. At Whitwell a fissile rock was formerly worked in a delph on Bakestone Moor close to the village (Fig 97) (Leo Godlewski, personal communication). There are in fact, two Magnesian Limestones. That from near Bolsover is much finer grained than the Whitwell stone. The diagenetic alteration [35] in the latter is also much greater and it contains no sand grains.

Market

There is a substantial market in the South Pennines which extends into the surrounding counties, especially north-east Cheshire. The 1996 South Pennine study (Hughes 1996) concluded that, in total, this amounted to about 2000 tonnes a year. It also included a partial assessment of the numbers of stone-roofed buildings in the region but it was not possible to analyse these in relation to each of the specific regional types. The least common type, a rock in the Magnesian Limestone, appears to exist on only three remaining roofs. To a large extent, all the others including the dominant type, Cracken Edge, are frequently replaced with reclaimed Yorkstone type.

The potential exists for this market to be segmented for each of the local types. However, at present, this is largely impractical except for the two most distinctive types; Freebirch and the Magnesian Limestone, and these are currently entirely dependent on reclaimed slates. The provision of a supply of each type, except perhaps the Magnesian Limestone, is the objective of the local authorities and some progress is being made towards this. Good communication between the conservation bodies and potential manufacturers is establishing a clearer understanding of the essential characteristics of each type.

Production

The industry in the region appears to have died out in the 1950s. The impact of this on the conservation of buildings was not felt immediately but problems gradually became more serious. By the 1990s the most significant source of stone slates for re-roofing, both new and reclaimed, was from Yorkshire and these were not a good match for the majority of the regional types. Since 1994, there has been a continuous effort to try to reverse the situation and by 1998 it was thought that this effort was making good progress. However, the several attempts to reopen old delves have suffered a variety of setbacks, including ill-founded objections from the public, excessively restrictive legislation and lapses of enthusiasm by the prospective delvers.

The application to open a small delve close to the important Magnum Bonum Quarry near Holmfirth was rejected by the Peak Park because of objections from English Nature. The probable consequence of protecting this one hectare of designated land is that more stone slates will be imported from India with far greater environmental impact.

This region is predominately rural and it has been recognised that there is a lack of people with industrial and commercial experience willing to undertake delving of stone slates. This contrasts strongly with the situation in the industrial areas of south Yorkshire. However, the farming community has the required attributes, skills and often the reserves of fissile rock, and an effort directed at farmers to persuade them to diversify into this field is receiving some positive responses.

In spite of the difficulties encountered there have been some successes. The long-established small delph at Fulwood Booth, and one near Bretton now have planning permission. The recently established Conservation Area Partnership Scheme in the Peak Park has been successful in returning two buildings to stone slate: one from Welsh slate and the other from concrete tiles.

Since the 1996 study (Hughes 1996) provided a structure within which to define the visual characteristics required for particular roofs, some of the Yorkshire quarries have begun to try to produce appropriate products. It is very encouraging to see an increased sophistication entering the market in this respect but it is not always the case. Some delph operators who are sawing to size are unwilling to dress the edges of slates in a satisfactory way.

Both the successes and failures experienced in trying to re-establish delves in this region point up the importance of a sustained effort on the part of the proponents. Entrepreneurs are hard to find, the mineral planning process is sometimes difficult and slow and public objections, though often ill-informed, can result in proposals being withdrawn. Such progress as has been made, has only been as the result of the long-term efforts of the Peak Park and the Derbyshire County and District building conservation officers. Details of delves supplying slate in Derbyshire and the Peak Park are listed in Annex A.

Other sources

Foreign sources

Several companies have started importing split sandstone from India. Currently the bulk of their sales are for flagstone flooring but one contract for the roofing of new-build houses in Glossop has been completed recently. The sizes supplied were 22 to 12 inches (560 to 300 mm) because the original specification had been for Bradstone concrete imitations. Normally, the importer would prefer to supply larger sizes, 36 to 20 inches (915 to 510 mm), for economy of coverage. They have experienced some difficulty in producing correct edge details mainly because they are working without a clear understanding of what is required.

If supplies of this sort are to become a part of the market, and the indications are that sales are increasing quite rapidly in Yorkshire, there is a need to establish a clear basis under which they will be acceptable to conservation bodies and an agreed form of testing to confirm their technical suitability. Conservation policies will normally encourage indigenous sources for reasons of sustainability and local distinctiveness. However, where it is clearly impossible to source a new local supply, even by a temporary delph, imported substitutes may be appropriate if they are of a similar range of sizes, colours, surface textures and mineral content, and are durable. To confirm this, at least a petrographic analysis by a geologist will be needed.

Reclaimed slates

The majority of roofs, both new and repaired, use new or reclaimed material from outside the region, mainly Yorkstone type which is a poor match for many local roofs.

Summary

From a very sorry situation a few years ago, when there was no source of new local stone slate of any type, some progress has been made towards re-establishing delves within a structure which should ensure that appropriate products are used in most situations. Where local supplies are still unavailable, suppliers from outside the region are becoming more responsive to the specific requirements of the South Pennines.

The cost of stone slates will always be a strong determinant of how many roofs are conserved and whether new roofs will be built with stone or with imitations. The entry of new, cheaper slates from other countries may be a competitive alternative to concrete. In so far as the weathered appearance will be more authentic than that of concrete, this is desirable. However, the presence of cheaper imports will not be beneficial to local production and may, in the long run, take some of the conservation market.

LANCASHIRE

> A whitish grey slate is raised in a great number of places to the south of the Lancaster Sands; but the quarries are seldom wrought to any great extent, as it is very heavy, and chiefly used for inferior kinds of buildings.
>
> There are several quarries in the parish of Milling, near Lancaster, where they get excellent slate of this kind. There is also some raised in Wyersdale, about Longridge, and near Blackburn. Mr Walton of Marsden-hall, works quarries of this sort, and there are still others near Colne. They are likewise met with more to the south-east and west, as about Todmorden, Rochdale, and in the neighbourhood of Chorley, Bolton and at Up-Holland, near Wigan, as well as in several other situations. (Dickson and Stevenson 1815, 78)

Geology

In describing a whitish grey slate, as opposed to a whitish-grey slate, Dickson was pointing to the different roofing materials produced in Lancashire; the blue or metamorphic slates quarried to the north of Morecambe Bay (now part of Cumbria) which are grey-coloured and the 'grey', that is, stone, slates from the fells and moors in the east and south of the county. He also remarked on the variable nature of the rock and the products which could be obtained from it:

> There is a fine bed of this kind of stone at Ellel Grange, on the property of Edward Rigby, Esq. It is also met with in Wyersdale and Claughton; and at Longridge, there are quarries of it, of very good quality. More to the south-east, it is likewise found in such a number of places, that it would be tedious to mention them. In the high range of hills between Bolton and Chorley, good stone of this sort is often met with also in the vicinity of Wigan, and near

Ormskirk. At Scarisbrick there is a quarry of excellent freestone of a close structure, on the property of the late T. Ecclestone, Esq. In some parts, the bed or delph is capable of being separated into thin pieces, proper for flags, and also occasionally for slates. (Dickson and Stevenson 1815, 79)

Carboniferous

Stone slates have been produced from two major divisions of the Silesian, or Upper Carboniferous: the Namurian or Millstone Grit Series of older terminology, and the Lower and Middle Coal Measures of Westphalian A and B age (Table 21). The exploitation of fissile sandstones in this region (indeed across the whole of Lancashire and Yorkshire) falls into two groups: the many, often large delves that served the markets of the industrial regions from Liverpool to York (those which Dickson found in 'such a number of places, that it would be tedious to mention them') and the smaller, more dispersed quarries of the rural areas to the north. The character of the latter group is typical of almost all the other stone slate producing areas of England: small and serving a small, local market. The contrast with the scale of the stone industry across the Lancashire-Yorkshire industrial belt could hardly be more dramatic. Here, the industry grew to supply every type of sandstone product to the rapidly developing towns and cities during the first industrial revolution. The production of stone slates was, of course, intimately linked with flagstones, but supplied markets within the region whereas flags were distributed on a national scale for flooring factories, paving streets etc. The sources in the industrial belt, in Rossendale and South Yorkshire, are described below. However, it should not be overlooked that there were very many small, local or temporary delves scattered throughout the region wherever Carboniferous stone was accessible, which have gone unrecorded.

North Lancashire

The Carboniferous sandstones of the Pennines extend westwards into Lancashire in the region north of the Calder Valley to Morecambe Bay. Stone slates from the Millstone Grit were produced from an early age and delves south of Lancaster, at Galgate and Dolphinholme (Dickson's Ellel quarry is here), and around Longridge became well known sources supplying over a large area. The sparsely populated area has also used stone slates from many small delves serving a purely local market. Examples of these roofs can be seen along the A683 from Lancaster to Kirkby Lonsdale and in Ingleton. Several quarries still operate in these areas but do not list stone roofing in their product range.

Ormskirk to Wigan

Elland Flags. [sic] These flagstones extend over a wide area in West Yorkshire and are by far the most important source of Carboniferous slates. As the Upholland or Rochdale Flags they are found in East Lancashire and their distribution extends into Derbyshire, as the Wingfield Flags and into North Staffordshire, where they are known as the Alton Rock. (Walton 1941, 66) [36]

Table 21. Sources of stone slates in Lancashire.

<table>
<tr><th colspan="4">Age</th><th colspan="2">Rock units</th></tr>
<tr><td rowspan="20">Upper Carboniferous (Silesian)</td><td>Series</td><td colspan="2">Stage</td><td colspan="2"></td></tr>
<tr><td rowspan="6">Westphalian</td><td>D</td><td>Upper Coal Measures</td><td colspan="2"></td></tr>
<tr><td>C</td><td>Upper and part of the Middle Coal Measures</td><td colspan="2"></td></tr>
<tr><td>B</td><td>Middle Coal Measures</td><td colspan="2"></td></tr>
<tr><td rowspan="2">A</td><td rowspan="2">Lower Coal Measures</td><td>Dynley Knoll Flags</td><td>= Clough Head Rock
= Deerplay Hill Flagrock
= Heald Flagrock
= Old Lawrence Rock</td></tr>
<tr><td>Crutchman Sandstone</td><td>= Milnrow Sst
= Tooter Hill Flagrock?</td></tr>
<tr><td rowspan="14">Namurian</td><td rowspan="3">G_1</td><td rowspan="3">Yeadonian – Millstone Grit 'Series'</td><td colspan="2">Rough Rock</td></tr>
<tr><td colspan="2">Rough Rock Flags / U Haslingden Flags</td></tr>
<tr><td colspan="2">L Haslingden Flags</td></tr>
<tr><td rowspan="3">R_2</td><td rowspan="3">Marsdenian</td><td colspan="2">Hazel Greave Grit</td></tr>
<tr><td colspan="2">Helmshore Grit</td></tr>
<tr><td colspan="2">Revidge Grit / Gorpley Grit / Fletcher Bank Grit</td></tr>
<tr><td rowspan="4">R_1</td><td rowspan="4">Kinderscoutian</td><td colspan="2">U Kinderscout Grit</td></tr>
<tr><td colspan="2">L Kinderscout Grit</td></tr>
<tr><td colspan="2">Todmordon Grit</td></tr>
<tr><td colspan="2">Cobden Sandstone</td></tr>
<tr><td>H_2</td><td>Alportian</td><td colspan="2"></td></tr>
<tr><td>H_1</td><td>Chokerian</td><td colspan="2"></td></tr>
<tr><td>E_2</td><td>Arnsbergian</td><td colspan="2"></td></tr>
<tr><td>E_1</td><td>Pendleian</td><td colspan="2">Wilpshire Grit</td></tr>
</table>

At Upholland, Orrell, Billinge, Scarisbrick and, no doubt, at other localities in the area, rocks of Namurian age (misnamed as Elland Flags by Walton) were sufficiently important sources of roofing to have been mentioned in several early references. In later documents they appear to have been eclipsed in importance, at least in later commentators' minds, by the scale of the industry to the east around Rossendale.

> These beds form the lower part of the Coal-Measures and consist of a series of micaceous flagstones, shales and thin beds of coal (Mountain Mines) attaining a total thickness of 1800 ft at Upholland. The base rests on the 'Rough Rock' of the Millstone Grit as shown at Grimshaw Delf ... The sandstones of the series ... are generally evenly bedded, micaceous, ripple-marked, and exhibit sun-cracks, tracks of Annelids and perhaps molluscs. They are worked extensively for flagstones and roofing slates, which split along planes formed of the small flakes of mica. The principal quarries are at Moss Bank, Billinge Hill, Rainford Delf, Bispham, Up Holland, Roby Mill and Tawd Quarry in Latham Park. (Hull 1860, 8–10)

Although there are many stone roofs to be seen in the area, no delves are known to be producing roofing in this area at present.

Rossendale

> Near Haslingden is Cold-Hutch-Bank, under a hill from which the finest flags and slate are quarried out. (Aikin 1795, 278)

The Haslingden Flags, the Dyneley Knoll Flags, the Old Lawrence Rock and the Crutchman Sandstone (which is the equivalent of the Milnrow Sandstone), have all produced roofing and, near Darwen, the lower part of the Crutchman, the Darwen Flags, has been extensively mined and quarried. Sources are listed in Table 22. Understandably, in view of the very large flagstone industry in this area, most references are to this product but most flagstone delves would have inevitably produced thinner stone suitable for roofing.

Haslingden

It was at Hutchbank and around Haslingden Town that the name Haslingden Flag was specifically applied because of its 'great value for flags, kerbs and sets' (Bolton 1890, 71). In scale of production, the most important rocks in this area were the Upper and Lower Haslingden Flags. 'The Lower Flags attain their maximum development along the Rossendale Valley, between Haslingden and Bacup' (Wright *et al* 1927, 22) Towards the north-west they continue as far as Blackburn, 'where the size of the quarries along the outcrop shows that a great mass of stone has been obtained in the past' (ibid), but in other directions they thin and disappear. The Upper Flags are more variable than the Lower but have been worked 'on the south side of Rossendale and maintain their character as economic flagstones in the Whitworth Valley District where they consist of 75 to 100 ft of well-bedded fine-grained micaceous sandstones' (ibid). To the west of Haslingden the Upper Flags were worked at Haslingden Grane to Pickup Bank and to the south at Wayoh Fold, Edge Fold and Edgworth.

Hutch Bank on the west side of Haslingden seems to have been one of the earliest sources of stone slates. In a survey of 1662 it was asked 'what Slatemines are within the Hundred of Blakeburnshire ... and who sels the Slatestones of Haslingden, Grane Post and the Pikelaws'. In answer it was explained that the workings had been leased to Edward Kippax, but were now 'almost worne out and little to bee gotten saveing what remains in the Coppihold lands' (lands rented from the lord of the manor) 'of one Lawrence Heape and Richard Rothwell in the Graine in Haslingden' (Baldwin 1998, 3). The slate pits were concentrated along the old main road, Stoney Rakes, about half a mile west of the village and the remains of the many small pits dating from this period can be seen along the B6232. To the east are the delves of Haslingden Grane and to the north there was another group around Baxenden. [37] These too may well have been quarried out at an early date. In 1439 rents were being drawn by the Receiver of the Honour from Baxstondelf for stone slate. On the high ground to the west of Haslingden stone slates were produced in delves around Top of Slate and further north at the Crutchman's Quarry working the Crutchman Sandstone which was so famous as a source of fissile rock that it became known as the Flag and Slate Rock. Further north the large quarry at Hameldon Scouts worked the Old Lawrence Rock 'which was known locally as the Hameldon Scout or Coppice Sandstone ... greenish-tinted, yellowish flag, which frequently bears a close resemblance to the Haslingden Flags' (Wright *et al* 1927, 155).

Balladen to Whitworth

> 1552 There were 'mynes or delfes on the Kings waste ground at Balladen wherein slate stones were got'. They were scarcely worth letting by the year for 20/- because 'there is slate enough to be had near adjoining on other men's land and because it is a wylde savage countrey farre from any habitacion'. (Baldwin 1998, 3)

Moving eastwards from Haslingden, the next major group of delves are found on the high ground to the south of the Rossendale Valley roughly bounded by the A56, A681 and the A671 roads.

> The larger of the quarries were Bankside, Deerplay, Heald and Rochdale Road quarries. In Stacksteads Frost Holes, Greens Clough, Law Head and Lee Delf were among the larger quarries. Bacup and Stacksteads form part of the Central Plateau, covering an area from Sharneyford to Haslingden and from south Burnley to the south of Ramsbottom.
>
> The area contains many forms of rock with names: Old Lawrence, Accrington Mud Stone, Dyneley Knoll Flags, Milnrow Sandstone, Bullion Mine Rock, Ganister Rock, Woodhead Hill Rock, Rough Rock, Upper Haslingden Flags and Lower Haslingden Flags (Taylor 2002).

Table 22. Sources and historical production centres of stone slates in Rossendale (after Bolton).

Formation	Horizon		Thickness in feet	Quarry locations	Comments
First Grit or Rough Rock	Deerplay Hill Flagrock		10	on the summit of Deerplay Hill, Longshaw, Easden Wood	Fine-grained grey flagrock with strong laminations, the surface of the flags being much ripple-marked.
	Black shale		60		
	Heald Flagrock or Old Lawrence	Upper HF	5	along the hillside from Sharneyford to Heald	In past times these beds have been much worked, the rock readily splitting into thin flags of 1 to 2 inches in thickness, which were used for roofing purposes during a lengthened period antecedent to that of the introduction of Welsh slate.
		Shale	12		
		Lower HF	12		
	Shale		30		
	Clough Head Rock	Flaggy sandstone	6	on the top of Little Tooter Hill	This bed is formed of a fine-grained yellow rock, flaggy towards the top, but becoming massive below. It forms an excellent building stone.
		Micaceous sandy shale	3		
		Broken flags	2–3		
		Massive rock	15		
	Shale		120		
	Tooter Hill Flagrock		30–60	on the sides of Tooter Hill	At Tooter Hill, a soft, crumbling, coarse sandstone, which is much false-bedded. Where it is well developed, as at South Grain Coal-pit in Dulesgate, and at Warmden, near Accrington, it forms a good paving stone.
	Blue & yellow sandstone		11		
	Shales, coals, ganister and fireclays		210–240		
	Woodhead Hill Rock		24	along the foot of the slope to Ash Cliffe, Bacup, Britannia Mill, Dulesgate, Nut Mill, Todmorden Road	
	Brown & black shales, coal and fireclay		88		
	Upper Rough Rock		15		
	Black shale		1		
	Sandrock mine		1		
	Fireclay		2–5		
	Lower Rough Rock		16–60	Bank House Lane, Sheep House Clough, Lee Height	
	Dark shale		30		
Second Grit or Haslingden Flags	Hard Sand Rock		44	Crag and Britannia	It is an extremely hard stone and very much used.
	Brown and Drab Shale		130		
	Black Shale		0		
	Brown Shale		20		
	Fine-grained Sandstone		27	large quarries opened into it at Hutchbank and at Hablingdon	To this rock more especially the title of 'Haslingden Flag' has been given. It is of Haslingden great value for flags, kerbstones and setts.
	Brown Shale		2–33		
	Strong Sandstones			Elton Bank, Edenfield and Brow Edge quarry, Great Height	It is extensively quarried for flags and ashlar at these sites.
Third Grit	Hard Brown Shale		98		
	Shales and coal		77		
	Sandrock		20		Very hard and fine-grained rock
	Shales, coals, Blackrock & fireclay		165		
	Flagrock		15–20	Balladen and Ravenshore	Flaggy sandstone
	False-bedded Sandrock		5		
	Shale, coal and fireclay		7		
	Sandrock		60	Green's (Beater's) Clough	Very false-bedded, and but loosely compacted
	Shale and coal		10		The shale strong and raggy
Fourth Grit or Kinder Scout Rocks	Sandrock		27		Both (Sandrock) beds are extensively quarried for ashlar and the better forms of building stone
	Shale		5		
	Sandrock		42		

In spite of the poor commercial prospects for stone slate manufacture in 1552, by 1620 they were of sufficient value for 'Edward Rawstone the elder of Lumb Hall to grant to his daughter-in-law Jane "all suche ... walestone, freeston, and slate, as ly on my Lanndes called Hancocke and alsoe all those walestones wich I have bought from Mr Rawstorne and another person, of slate standinge in Balliden Hey"' (Baldwin 1998, 3).

The delves throughout this area developed to become substantial workings. Depending on the exposure of suitable rock and in order to minimize overburden removal, they operated either on flat land, often at a break in a slope, or into a scarp face. One example of the former at Horncliffe Quarry was the subject of a 1784 lease in which all the slate mines [38] were valued at an annual rent of £18 18s. Brow Edge Quarry is an example in a scarp slope. In the 1780s, slate mines were operated at Foe or Fall Edge and Tottington Upper End and at Higher Booths and Crest Height.

Many of these delves were in the Upper and Lower Haslingden Flags which extend as far as the Whitworth Valley in an eastwards direction and southwards around Dearden Moor to Scout Moor Quarries and Fecit End Delf. Although at Horncliffe Quarry there is 100 ft (30.5 m) thickness of workable Lower Flags which is maintained through the many large delves in an easterly direction to Cragg Quarry and the Greens Moor and Lee Quarries, they thin to about 6ft (1.8 m) at their southern extent.

The Upper Flags were also extensively quarried to the south of Rossendale Valley (on the north side they are poor and thin) and around Whitworth and at Fecit Quarry are as much as 100ft thick (30.5 m). To the north of Rawtenstall the Upper Flags increase in thickness and strength and were quarried on Bonfire Hill to the east of Crawshawbooth.

Heald Moor

> almost all the old houses in Rossendale were roofed with tiles from these quarries ... (anon 1891)

While it might be an exaggeration to imply that the other delves in the Rossendale Valley had contributed little to the roofing of houses, the delves of Heald Moor are certainly old and important. Indeed, the name Heald indicates that roofing was a significant product of the Moorside delves: healing or heleing is an old word for roofing or slating and is still in use in some regions. Maps of the area dating from as early as 1578 show 'sclatte pittes' and a number of seventeenth-century deeds and contracts refer to slate and flags. Throughout the nineteenth century Lord James and Company, variously of Held, Heald Delph, Heald Top and Clough Head, appears in a number of documents as stone merchants and suppliers of pierpoints, slate, flags, landings, cornishes, fire pieces, hearth stones, wall stones and slop stones, indicating the flaggy nature of the local stone and the products made from it.

Stretching from around Holden Gate through Slate Pit Hill and Tooter Hill to Deerplay Hill the delves are now largely overgrown. They mainly worked the Dynley Knoll Flags, but extended into the Darwen Flags (or their locally named equivalent) and the Milnrow or Crutchman Sandstone.

Bolton records the Dynley Knoll Flags, known locally as the Tooter Hill Flagrock, being worked on the same named hill and, although he describes it as soft and crumbling at this location, he says that elsewhere, in Dulesgate, and at Warmden near Accrington, it forms a good paving stone. On the summit of Deerplay Hill, at Longshaw and Easden Wood this is a 'fine-grained grey flagrock with strong laminations, the surface of the flags being much ripple-marked'. Also, speaking of the Old Lawrence Rock or Upper and Lower Heald Flagrock, he describes how 'in past times, these beds have been much worked, the rock readily splitting into thin flags of 1 to 2 inches in thickness, which were used for roofing purposes during a lengthened period antecedent to that of the introduction of Welsh slate. The refuse tips from these old quarries extend along the hillside from Sharneyford to Heald' (Bolton 1890, 91). At Clough Head the Clough Head Rock, a fine-grained yellow rock, flaggy towards the top becoming massive below, was worked.

North of this area, from Blackburn to Colne and onwards into south Yorkshire, there were other delves which produced roofing on a smaller scale.

Market

Together with the markets in the Yorkshire industrial belt and within the Yorkshire Dales National Park and the surrounding rural area there is a substantial demand for sandstone slates in the region as a whole. Within this region, East Lancashire has a significant demand.

Currently, there is little attempt to distinguish between the different stone slate types, even between Gritstones and Coal Measures stones (Namurian and Westphalian) although the better informed (local) architectural practices do try to use specific types where possible. The failure to specify or source specific rocks is largely a consequence of the limited supply but, with the increase in the number of quarries working these sandstones, this situation may improve in future.

Production

Crutchman's Quarry, a delph in the Crutchman's stone, was opened south-east of Accrington but because of a mix-up in gaining permission to work it had to close down. This is a pity because this stone had such a good reputation that it became known, in its heyday, as the Flag and Slate Rock. The only other source of new stone slates known in the region is from the Deerplay quarry which works the Old Lawrence Rock or Dyneley Knoll Flag. Details of suppliers in the Lancashire region are given in Annex A.

Other sources

With a few exceptions the majority of roof renovations and the few new roofs which are being built in the region

Figure 98. Slates from the Elland Flags, Bradford (© Terry Hughes).

Figure 99. At the Rand and Askwith quarry in the Elland Flags, the line of the old stone slate mines can be seen high in the quarry wall where the gallery roof was supported with stone walls (© Terry Hughes).

are being supplied from reclamation sources, from quarries in Yorkshire or, increasingly, from India.

Summary

Stone roofing has a long history in Lancashire in both the industrial towns and cities and the rural uplands of the north. Although one delph in the region can supply authentic local stone, most stone roofing relies on supplies from neighbouring counties or from foreign imports. Often the resulting roofs do not have an authentic appearance.

YORKSHIRE

> Quarrying of the Elland Flags in the type area of Elland Edge and in the neighbouring areas of Raistrick, Hipperholme and Northowram, near Halifax, was a thriving industry at the end of the last century and until the First World War. It no doubt originated to provide thakstones, and the early quarrying operations were concentrated along the edges of the valleys, where the stone outcrops could be easily attained. By the seventeenth century definite quarries were excavated in most districts, proof of which is afforded by the fact that William Judson, James Dalton and Thomas Wood were each fined 6s 8d for not filling up nor fencing delves at Norwood Green. (Walton 1975, 45)

During the eighteenth and nineteenth centuries a very large industry developed in West Yorkshire exploiting the sandstones of the Millstone Grit (Namurian) and the Coal Measures (Westphalian) for masonry, flagging and for roofing, known locally as thackstones (Fig 98). Generally the delves worked exposures on the steep valley sides, where there was less overburden to remove, but also extending into the plateaux beyond. There are many descriptions of the rows of cranes in quarry after quarry along the skylines of Elland Edge, Southowram and similar centres. The extent of the industry has been reviewed in Godwin (1984). The fissile rock was sometimes followed underground. At Rand and Asquith's Tuck Royd Quarry near Brighouse the underground workings can be seen high up in the wall of the modern quarry (Fig 99). Walton (1938) has described the early methods of working.

> From the records of the building of Rastrick Chapel in 1602 (Lister: 'Rastrick Chapel and School', Halifax Ant. Soc. Trans., 1905), it appears that landowners employed labourers to quarry the stone on their own land. In the Elland Edge district this ultimately developed into the 'master-taker' system, whereby the landowner let out the getting of the stone to a 'master-taker', who undertook to get the stone for a stipulated sum. He employed 'delvers', but he was often regarded as unscrupulous by both the landowner and the delvers. Frequently the overburden from a working was tipped on the top of unworked land and the whole declared to have been quarried. The 'master-taker' was also usually the local innkeeper, and, as he paid out the wages of his 'delvers' in his inn, a goodly proportion of their earnings was quickly returned to his coffers. This system survived until well into the nineteenth century.
>
> The early quarrymen, who worked the outcrop, took only the best stone, and, as this became worked out, galleries were driven into the hillside. The development of deep quarries and mines came only with the introduction of gins and steam cranes, and in Hipperholme the earliest shafts were sunk about 1860. The overburden removed when sinking a mine was hauled to the surface by means of a windlass and built into a massive retaining wall, known as a 'judd wall'. Behind this wall other waste material was tipped to form a level field in due course.
>
> When the shaft reached the required depth, galleries were run under the ground, as at Collyweston, although the galleries were much more extensive. The mining operations were controlled by the 'wooder', who had charge of the timbering, and in order to conserve timber he also determined the nature and direction of the galleries. Props were not used and the timbering was restricted to the roof, which was wedged by means of ash 'lids'. The mines took advantage of a thin layer of soft shale under-

Table 23. Sources and historical production centres of stone-slates in West Yorkshire.

Age		Rock unit	Locations	
	D			
	C			
	B			
Westphalian	A	Penistone Flags		
		Elland Flags	Southern Yorkshire, Halifax to Bradford	
		Gaisby Rock	Bradford	
		Soft Bed Flags		= Crawshaw Sst
Namurian	G_1	Rough Rock Flags	Halifax and Huddersfield	
	R_2	Barkisland Flags	Over a wide area	
		Beacon Hill Flags	Halifax	
		Scotland Flags / Readycon Dean Flags	Halifax	
		Pule Hill Grit	Meltham Moor Holmefirth Holme Valley	
		Huddersfield White Rock		

lying the stone, called the 'pricking'. Working by candle-light and in a height of only two to three feet, they removed the 'pricking', when the rock above was left to 'weight' and fall by gravity.

When the quarries, or 'delves', were shallow the blocks of stone were carried to the surface by 'huggers', who wore leather 'saddles', similar to those formerly worn by coal hawkers, and were capable of carrying loads up to eight hundredweights. The stone was carried up 'hugging ladders', having very broad rungs set close together, and as the 'hugger' had to hold the block of stone he could not use his hands to steady himself. Fatalities often occurred and there are records of ladders breaking under the strain and the unfortunate 'huggers' being killed.

'Hugging' continued until about 1870, although gins and hand-cranes, capable of lifting a ton, were used earlier in some quarries. One of the most primitive haulage devices was the 'Billy Wobble', which was merely a jib, over which passed a rope or chain into the quarry. The lower end of the rope was split into three parts, to two of which were attached iron rings and to the third a hook. The stone was loaded into a 'shim', which has a sloping front board and no sides. The rings were slipped over the 'strines' (handles) and the hook was attached to the 'tell' (wheel). A horse was harnessed to the free end of the rope and as it moved away, it hauled the loaded barrow to the surface. A similar method is still employed at Collyweston. Power cranes came into use about 1870, the first type being called by the quarrymen a 'monkey up stick' crane. Wire ropes were not used until about 1880.

Elland Flags in all forms were transported from the quarry on stone wagons, few of which have survived. An early example of such a wagon is depicted on Robert Parker's plan of Harrison Farm, Southowram. Later wagons were much sturdier. They had wooden axles with four plates at each end on which the massive wheels rotated. They required constant lubrication, and the carter invariably carried a horn of grease and a feather, with which to lubricate the wheels every mile or so. Such 'heavy' wagons quickly damaged the poor dirt roads, and stone tracks, consisting of parallel rows of thick flags, were laid to the quarries. Many of these tracks, deeply rutted, have survived. (Walton 1975, 45–7)

Whitehouse, in a report for the Yorkshire Dales National Park, also described in some detail the methods adopted in surface and underground workings.

Moorhouse investigated medieval stone slate quarry records in the region and excavated some quarries. One of these, at Thorner in Humberside, produced nine waggon-loads of 'sclatston' in 1465–6 from Carboniferous rock (Moorhouse 1990, 137).

Moorhouse's study also reached some important conclusions about the industry from the fourteenth to the seventeenth centuries and the use of stone roofing in the region. During the Middle Ages most buildings in West Yorkshire were timber-framed and he concluded that the 'overwhelming use of stone in the region was as a roof covering and to a lesser extent as wall panels set in the timber frames' (op cit, 126). A wider range of buildings were roofed with thackstones than has been generally thought [39] and 'by all levels of society' and 'on most types of building … hall, kitchen and oven are the most obvious but stables and barns are also mentioned fairly frequently' (op cit, 126–45). That they were highly valued is indicated by the price people were willing to pay for transport. Examples of carriage three times the cost of the slates are quoted.

Geology of the Carboniferous

West Yorkshire

Stone slate manufacture has mainly been in the Upper Namurian and Lower Westphalian. Although there are many different named rocks within the region which were exploited for flagging and roofing, the Elland Flags came to dominate the market, limiting the exploitation of even the high quality Rough Rock Flags (Walton 1941,

Figure 100. Between Halifax and Bradford there were major sources of stone slates at Elland Edge, Raistrick, Hipperholme and at Northowram (© Terry Hughes).

Figure 101. A roof of Elland Flags in Bradford (© Terry Hughes).

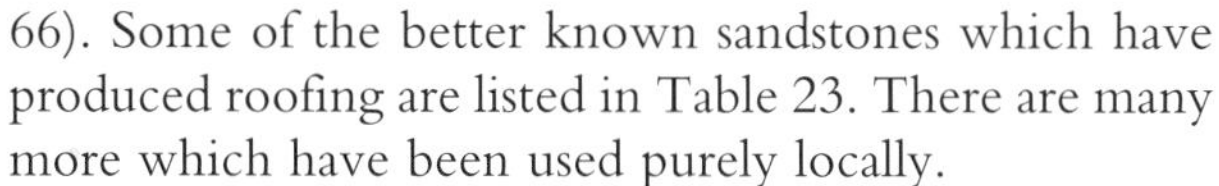

66). Some of the better known sandstones which have produced roofing are listed in Table 23. There are many more which have been used purely locally.

Early production extended throughout the region serving local markets but eventually, and as roads improved, some locations developed into regional production centres and for flagging they were of national importance. The most important, in the Lower Coal Measures, extended from Elland Edge through Rastrick, Southowram, Hipperholme and Northowram to Bradford (Figs 100 and 101) and Idle. Around Halifax and Huddersfield, the Rough Rock Flags of the Millstone Grit were extensively worked and there were important delves north of Bradford where Moorhouse recorded medieval workings at Manningham, Calverly, Pudsey, Bramley and Woodhouse. In the south of the county there were important delves around Holmfirth (see Fig 89), the Holme Valley and Meltham Moor. At Harden Clough a very large delph working the Rough Rock and Rough Rock Flags (and the Rough Rock Coal) supplied flags as far away as Manchester (Fig 102). The size and quality of the rock was so good that the delph became known as the Magnum Bonum. (There is a further description of delving in this area in the Derbyshire and Peak Park section.) In the north-east of the region the Millstone Grit continues as a source of roofing across Airedale and Wharfedale towards Harrogate. In the Middle Coal Measures, the early delves near Wakefield, at Alverthorpe, Snydale and Featherstone, for example (Moorhouse 1990), seem to have fallen out of use because of poorer durability. (It should not be inferred from this that the Coal Measure sandstones are consistently inferior to the Millstone Grit.)

> The fissile sandstones, flags and limestones, which occur at several horizons in these series, have been widely employed throughout northern England, but there were no well established centres, apart from those near Leyburn in Wensleydale. (Walton 1975, 39)

Walton went on to say that 'quarrying has been sporadic and on a small scale to meet the demands of local

Figure 102. At Harden Clough south of Holmefirth the Rough Rock and Rough Rock Flags produced huge slabs, sometimes called covers, and supplied all the major towns of the region. The size of the slates and quality of the rock was so good the delves became known as the Magnum Bonum Quarry, Ne Plus Ultra Quarry and Nonpareil Quarry (© Terry Hughes).

requirements and no slates are now being quarried' (ibid, 39). Happily, some stone slates are being delved today in Wensleydale and at Barnard Castle in County Durham and in Northumberland, and it is possible that, with the encouragement of the Yorkshire Dales National Park, other delves will open in the future. Although Walton regarded the North Yorkshire industry as small it was in fact quite substantial and there was a large export of flagstone throughout England at one time.

Yorkshire Dales

In 1989 the Yorkshire Dales National Park Authority commissioned a report from J D Carlisle of Whitehouse Services on the historical and potential modern sources of stone slates in the region, and the size of the local market defined as Richmondshire, Teesdale, Craven and south-east Lakeland (Fig 103). The following section draws on that work.

Stone slates from both the Lower and Upper Carboniferous have been used for roofing. Carlisle examined 51 old quarries and some of the localities and their position in the succession relative to local limestones are listed in

Figure 103. In the Yorkshire Dales there were many small quarries, most supplying just the immediate area. This roof in Dent has crowsteps in the chimneys which act as a damp course and to throw water away from the roof abutments (© Terry Hughes).

Figure 104. A Horton Flag roof near Ingleton (© Terry Hughes).

Table 24. Of the 51 quarries reviewed, 12 were discounted as not having suitable rock and 27 were considered to be impractical for a variety of reasons. Of the remainder, seven were good prospects for further investigation and five were held in reserve. These are included in Table 25.

In Ribblesdale the Silurian rocks exposed in the inlier in the Helwith Bridge area have been a source of stone slates in Arcow Quarry (usually identified as the Horton Flags) but are not delved at present (Fig 104). Both the Helwith Bridge Quarry (David Jefferson, personal communication) and Dry Rigg Quarry (Jonathon Ratter, personal communication) are believed to have some fissile rock.

> In Coverdale one of the smaller Dales which penetrate the western moorlands a stone slate is procured. The stone out of which it is riven lies on a level at the water at the foot of the hill on the western side of the dale and is supposed to extend far under it in a stratum not more than three feet thick; this affords a tolerably good, but heavy covering, as far as the expense of land carriage will admit of the use of it. A similar slate is also got at Pen-Hill, lying between this dale and Wensleydale.

Table 24. Some old sources of stone slates and production centres in North Yorkshire (after Carlisle).

Age				Rock unit	Location or quarry	
Silesian	Namurian	E2	Arnsbergian	Grindstone Limestone		
				Unnamed Sandstone	Shipley Banks	NZ018207
				Fossil Sandstone		
				Lad Gill Limestone		
		E1	Pendleian	Tan Hill Grit	Hill Top	NY856018
				Mirk Fell Ganister	Bowes Fell	NY9716
Dinantian		D2	Brigantian	Undersett Limestone		
				Unnamed sandstone	Hudson's Quarry Oswald's High Wood Witton Steeps	SE027865 SE036876 SE066867
				Five Yard Limestone		
				Unnamed sandstone	Brignall Washton Gayles Swaledale, Arkengathdale, the north side of Wensleydale from Garsdale and Cotterside to Wensley and Simonstone	NZ070123 NZ149057 NZ125073 SD750893 SD827927 SE093897 SD872918
				Middle Limestone		
				Unnamed sandstone	Thwaite Quarry Burtersett South side of Wensleydale at Penhill (Low Dove Scar), Hudson's Quarry, Gayle Bank	SD975891 SD892981 SE038863 SE072865 SE091886
Silurian			Ludlow Series	Horton Flags	Horton in Ribblesdale	Around SD805700

Table 25. Potential or actual sources of stone slates in the Richmondshire region (after Carlisle).

Quarry	Location	
High Pike, Dent	SD719827	
Cotterside	SD827927	
East Shaw and Scar Head	SD870884	
Stags Fell	SD873921	
Seavy, Old and Raygill	SD896887	
Penhill Park	SE060870	
Craneshaw Bottom	SE063938	
Gilbert Scar, West Grafton	SE064834	
Long Slack, Melmerby Moor	SE066864	
Gayle Bank	SE091886	
Hungry Hill	SE097888	
Hill Top, Keld	NY856018	operating
Bail Hill	NY970223	
Shipley Banks, Cotherstone	NZ018207	operating

> A slate somewhat resembling that which is usually called Westmorland slate, but of a coarser texture, and of a more colourful colour, is found in Swaledale but the use of it does not extend far beyond the place which produces it. (Tuke 1800, 20)

The different sources have produced quite a wide variety of stone slates within the region. More work needs to be done on this aspect of the region's conservation of stone roofs but the following brings together some of what has been recorded in the field and by Walton, Carlisle and others.

All sources have produced slates of roughly the same size but some, from the Burtersett area mainly, are thinner than is usual for Pennine sandstones and the Horton Flags are quite thick. The exception to the general run of sizes is those from Coverdale which are usually large.

Generally the slate's texture is grainy rather than smooth like the finer grained Coal Measures stones and some have a rippled surface. The colour varies over the whole region from grey to light buff and may include iron staining (Fig 105). However, after being on the roof for some time they all tend to weather to similar colours. Edge dressings vary: in Swaledale the tradition is for the bevels on the long edges to face downwards but at the tail they face up. For this reason (and others) reused slates should not be turned over. Further north the bevels on the long edges are sometimes opposite, that is upward and downward, so that when laid they overlap. Edge treatments are often the personal preference of the manufacturer or roofer but where there is a tradition for a particular type care should be taken to include an appropriate item in roof specifications.

Walton reported that Coverdale slates were the region's best and that 'Carperby were of poor quality, Burtersetts broke and those from Stags Fell, north of Hawes were crooked' (Walton 1975, 39). No doubt the poor durability of the Burtersetts was offset by the economic advantage of their thinness (Fig 106). If the Stags Fell slates were crooked, and this could have been a temporary situation, it does not seem to have limited production, as Carlisle records that it is a site known 'to have produced good slates in large quantities' (Carlisle 1990, **1** 3). Carlisle also recorded, from his list of quarries

Figure 105. Millstone Grit slates at Hill Top Quarry, Swaledale, which works the Tan Hill Grit (© Terry Hughes).

worthy of further investigation, the following as having good material: Hill Top, East Bail Hill, Penhill, Gilbert Scar, Gayle Bank and Hungry Hill.

Jurassic

Brandsby

> In the Howardian Hills the Grey Limestone Series is much more flaggy than it is over the northern region and, having been quarried for a great number of years at Brandsby as a material for mending roads, is better known in this district as the 'Brandsby Roadstone'.
>
> It is a hard siliceous limestone splitting up into large slabs, in fact some of the beds are so fissile as to afford roofing slates, for which purpose it was largely used in former years. (Fox-Strangeways 1892, 248)

The low land around the North York Moors is the northern limit of the Jurassic. Fittingly, it does not disappear into the North Sea without producing two more stone slates. In the area around Brandsby, a small number of roofs remain in what must at one time have been an island of limestone slates in a sea of clay tiles (Fig 107).

Figure 106. Swaledale. Thin slates, probably from Burtersett delph (© Terry Hughes).

Figure 107. Around Brandsby the local Jurassic limestone roofs have often been renewed with Carboniferous sandstone slates. Here the main slope is Brandsby stone but the left slope is sandstone (© Terry Hughes).

The first of these stone slates was obtained from what came to be known as the Brandsby Roadstone. It was an important stone at a time when all road metal was obtained from local quarries. However by 1892, Fox-Strangeways was already writing of 'the Brandsby Roadstone which was formerly extensively quarried in the western part of the Howardian Hills between Brandsby and Gilling; but which, since the introduction of railways, has been largely superseded by whinstone' (op cit, 463).

It lies near the top of the Middle Jurassic, in the Scarborough Formation (formerly the Grey Limestone) (Table 26). Fox-Strangeways *et al* (1886) described the locations of the outcrops.

> On the southern side of the north Coxwold fault the Grey Limestone is first seen at Thirkleby Barf, where it has been extensively quarried for road-metal. The beds here, which are harder and more massive than is usually the case, are divisible into two portions, the lower part consisting of hard beds of blue fossiliferous limestone, while above this are soft massive sandstones with casts of fossils, which become more flaggy in the upper part. These beds, which at Thirkleby Barf dip about 15 degrees to the north-east, to the east of Burtree House curve round and above Barf Hill, dip at about the same angle slightly to the west of north, so that the outcrop is extended in an easterly direction by Wildon Hill to Coxwold; at all of these places the rock is exposed in old quarries, but is not seen along the intermediate ground, probably owing to the thick covering of Boulder Clay lying in the hollows between. In the beck to the south of the railway at Coxwold these beds are seen dipping to the south-east at an angle of 10 degrees, being bent over towards the second large fault by which they are thrown up so as to outcrop only in the south-east corner of Newburgh Park, as outliers at The Mount and Park House .... Besides the main outcrop of the Grey Limestone there are several outliers. These occur as patches on the moors to the north, and fringing some of the hills in the neighbourhood of Kirkby Knowle and Kilburn to the south, the most northerly being that on Whorlton Moor. (Fox-Strangeways *et al* 1886, 4)

There are other outcrops in the north-west corner of the moors, on Snailsworth Moor and at Hood Hill near Kilburn.

A little further north, in the Hambleton Hills, another Jurassic limestone has been used for roofing: the Boltby Slate of the Middle Oolite. It is the Lower Limestone of Fox Strangeways *et al* (1886, 2); the present-day Hambleton Oolite Member of the Corallian Group (Richard Carr-Archer and Graham Lott, personal communication).

> In the district between Hawnby and Kepwick this limestone is fairly massive; but a much more barren rock than it was to the east; in fact, the paucity of fossils is remarkable; as we trace the rock to the south it becomes more flaggy, and on Boltby Moor is so fissile that some of the quarries have been marked on the Ordnance Map as Slate Quarries. (Fox-Strangeways *et al* 1886, 49)

Table 26. Jurassic stone slates in North Yorkshire. Formation names in Fox-Strangeways (1886) and modern equivalents.

Age	Fox-Strangeways			Modern	
Upper Jurassic	Middle Oolite	Upper Calcareous Grit			
		Upper Limestone and Coral Rag			
		Middle Calcareous Limestone			
		Lower Limestone	Boltby Slates	Hambleton Oolite Member of the Corallian Group	
		Lower Calcareous Grit			
		Oxford Clay			
		Kellaways Rock			
Middle Jurassic	Lower Oolite	Cornbrash			
		Upper Estuarine (Deltaic) Series		Scalby Formation	Long Nab Member
					Moor Grit
		Scarborough or Grey Limestone Series	Brandsby Slates	Scarborough Formation	Crinoid Grit, Brandsby Roadstone
		Middle and Lower Estuarine Series			Cloughton Formation
		Eller Beck Bed or Hydraulic Limestone		Eller Beck Formation	Gormire Limestone
		Lower Estuarine (Deltaic) Series		Saltwick Formation	
		Dogger		Dogger Formation	

Figure 108. Brandsby slates can be very rough. This example is one metre long (© Terry Hughes).

Figure 109. The roof of St Mary Magdalene at Over Silton shows the distinctly different Carboniferous sandstones at the eaves and the original Brandsby Slates above. It would be preferable to use a limestone slate from, say, the Cotswolds to make up the shortfall during re-roofing (© Terry Hughes).

The Flag Quarry Plantation on Scawton Moor to the south-east is also probably so called from the fissile character of the limestone. The outcrop of this limestone covers the summit of all the high ground to the south and west of Hawnby, and stretches away to the north as far as Arden Moor.

The stone slates of the region appear to be quite variable in appearance, ranging from reasonably flat and similar to a Forest Marble slate from the Cotswolds, to thick and uneven (Fig 108). This may simply be the difference between Brandsby and Boltby slates. Some roofs have a thin stone slate similar to Collyweston. This is assumed to be a substitute for the original stone but it has not been possible to confirm this.

Market

There is a large and constant demand for the Carboniferous stone slates of southern Yorkshire which extends into the neighbouring counties to the west and south. Yorkshire stone slate types are not always the most suitable choice for these neighbouring markets. Of course, there is little or no demand from the mills and factories of the past, indeed, reclamation from such buildings is still a significant part of the supply to this market. But today, stone slates are mainly used on new and refurbished houses and within conservation areas in the Yorkshire Dales National Park and on farm buildings.

Within the Dales region there is a significant demand for new stone slates largely generated by the very successful Barns and Walls Conservation Scheme. The scheme began in Swaledale in 1989 and was extended to include Littondale in 1994 and Wensleydale in 1996. Up to July 1999 the conservation of more than 300 barns and 22 km (13.64 miles) of walls had received funding totaling £1.77 million, from the Yorkshire Dales National Park and English Heritage, 25% each, and from the European Union, 50%.

The market for Brandsby or Boltby stone slates extends beyond the immediate area of production but even so, today it is very small and except for a few isolated examples lies within about a 16 km (10 mile) radius of the quarries. There will only be a very occasional need for stone slates, but the majority of the buildings with these roofs are important historic buildings and they should be conserved by using the most suitable means of obtaining an appropriate material.

Most roofs would originally have been covered entirely in Brandsby or Boltby stone but those stone slates which remain are concentrated on one or two slopes or at the eaves. Elsewhere they have changed to another covering entirely. Where an attempt is made to conserve the appearance of the buildings, the solution adopted has often been to use Gritstone slates from the Carboniferous to the west (Fig 109) or sometimes, it appears, Collyweston Slates. While Carboniferous stone slates are a better substitute than many other options, they do not closely resemble the rough limestones, nor do they support the same vegetation so the ultimate colour of the roof will be different as well. If supplies in the Cotswolds become adequate for local conservation, these would be the preferable substitute.

Wholly inappropriately, sandstone, which is not a traditional local roof type, imported from India, is being used on some new buildings in the area.

Clay pantiles are the dominant traditional material elsewhere in the region. Pantiles over Brandsby stone at the eaves is now a vernacular detail which should be conserved.

Production

A number of quarries in Yorkshire are working fissile rock and, although in most the production of stone slates is sporadic, many are in recognised roofing stones. Overall the supply is sufficiently reliable that with a reasonable lead time it is possible to obtain Gritstone or Coal Measures stone specifically.

Because Yorkstone flagging is such a popular material, stone which could be made into roofing, is often diverted to this end.

There is no production of Brandsby or Boltby stone slates today. Although delving of stone for walling and roofs probably continued after the demise of the roadstone, it is unlikely that there has been any serious production in the twentieth century.

The Yorkshire quarries sufficiently well organised to be able to take orders for roofing, subject to being in suitable rock, are listed in Annex A. There are many other small operations in Yorkshire which produce stone slates from time to time.

Other sources

Carboniferous

In the past reclaimed stone slates were available on a large scale, but the availability is now much reduced. Worryingly, dealers in reclaimed material are increasingly looking to imports from abroad, most notably from India, for supplies at a price close to that of reclaimed. These are being sold with no evidence of their durability once on the roof.

In the past at least one attempt was made to saw roofing thickness slabs from block stone, believed to be the Grenoside Grit. This was unsuccessful.

Other quarries in Lancashire, Durham and Northumberland can supply similar stone slates.

Jurassic

There are no sources of authentic limestone slates and there is no reservoir for reclamation, consequently, other Jurassic limestones, Carboniferous sandstones and imported sandstones from India are being used for conservation and for roofs of new buildings.

Summary

In West and North Yorkshire there are significant markets for a variety of Carboniferous stone slate types, sufficient to sustain a significant industry. The supply situation has improved slightly in recent years with the reopening of some old delves including some in adjacent counties. As a result it is now possible to obtain at least Gritstone and Coal Measure slates specifically. The prospects of all these delves are entirely dependent on a secure market and the recent advent of imports from India poses a serious threat to their future.

The limestone roofs of the Brandsby region are an important remnant of a vernacular roof type unique in the north of England. In this respect they should be conserved, but there are considerable difficulties standing in the way of this objective. The market is too small to sustain a delph and the only viable alternative would be to locate a delph with suitably fissile rock from which slates could be produced as required for individual buildings. Failing this, similar stone slates from the Cotswolds should be used in preference to the larger, flatter sandstones from the Pennines. An added conservation difficulty is that the local vernacular detailing may have been lost during later repairs and restorations.

DURHAM AND NORTHUMBERLAND

Carboniferous stone slates of Namurian (Millstone Grit) and Westphalian (Coal Measures) age were delved extensively in these two counties. Writing of Northumberland at the start of the nineteenth century, Bailey and Culley described the use of stone slates (they used the term freestones) but also acknowledged the presence of Scottish, metamorphic slates. The latter would probably be from the quarries of south-west Scotland and the Borders rather than the more well-known Highland slates. [40]

> Freestones of various kind abound in almost every part of the county and are applied to all the purposes of building. Many of the quarries afford tolerable slates for roofing and flaggs for floors at some of them. Excellent grindstones are got of which a great many are exported from Camas and Warkworth.
>
> Cottages. The materials used for building are stone and bricks but mostly the former. Straw used to be the universal covering but it is now nearly fallen into disuse and tile or slates substituted in its stead. Small dark blue slates from Scotland are the kind generally used here and are much superior to tiles although they are more expensive first, it is probable that in a few years they may be as cheap from the repairs tiles so frequently require especially when they are so ill manufactured. (Bailey and Culley 1805, 21–8)

A similar situation applied in County Durham although the metamorphic slate came from the west as well as Scotland.

> Slates for roofing of the freestone kind called grey slates are found in many places in the western parts of the county. Materials for artisan's and labourer's buildings ... coverings either Westmorland slates, Scotch slates or freestone slates, tiles, straw or ling the latter only in the western part of the county. (Bailey 1810, 44)

Westmorland slate per yd	5/-
Scotch slate	5/-
Freestone slate	2/9
Mason work of roofing with slates	1/- per yd

In spite of the inroads made by metamorphic slates from Scotland and Westmorland at that time, subsequently from Wales, and today from other countries, stone slates are still an important feature of the Pennine villages and farms.

Geology

Most of County Durham and Northumberland stand on rocks of Carboniferous age: the Millstone Grit and the Coal Measures, which form the Pennine uplands and, to the west and north, the Carboniferous Limestone. To the east lies the Magnesian Limestone. It appears that lime-

Figure 110. Newly laid Coal Measures sandstone from Ladycross Quarry, Northumberland, at Merrick Priory. The colours will darken quite rapidly producing a more even appearance. (© Terry Hughes).

stones have not been used for roofing but the sandstones were an important source of stone slates.

Although sandstones from both the Carboniferous periods have been quarried they are more alike than the rocks of the same ages in the southern Pennines. Speaking of Northumberland, Lebour (1878, 55) commented that the 'Millstone Grit [is] scarcely different to the Coal Measures – not nearly so coarse nor so thick as the Millstone Grit of Derbyshire. ..... [The] Millstone Grit is thin in the South and thinner in the North.' He added that 'in Durham the Millstone Grit is more important'.

Like South Yorkshire, sandstone quarrying developed to serve two distinct markets. Close to the industrial cities of the east a substantial industry developed. In contrast, in the upland, rural areas the overriding impression is of the remains of many small delves each associated with a small community or farmstead.

In the Coal Measures near Durham, Salzman (1952, 393) recorded, 'Works done during the time of John Fossor, Prior of Durham 1341–74. Also at Beaurepaire two new dairies, the whole sowtsyd of the wood and of the wall round the wood, and he discovered a quarry of sclatstane'.

The Coal Measures were also worked at Newcastle upon Tyne. Crane (1979) described an extensive industry, including the production of stone slates,

Figure 111. Carboniferous Coal Measures sandstone from Ladycross Quarry, Northumberland. Munsell scale 7.5 YR, centimetre scale (© Terry Hughes).

from as early as 1647. At first, quarrying took place close to the city but as building pushed outwards reliance was increasingly placed on workings further afield all round the city. Prominent amongst these were the quarries between Heddon and Stocksfield along the River Tyne. The Tyne Valley also provided a route to the city for stone slates from the group of Coal Measures delves in Slaley Forest. One of these, Ladycross Quarry, has continued to supply roofing throughout the region and as far afield as Herefordshire up to the present (Figs 110 and 111).

In 1827 the production of flagstones from the Newcastle region was important enough for them to noted by Mackenzie (1827, 715) in the list of exports from the port. Many of the delves producing flags would have been the source of the stone slates for the city's roofs. Often they were operated by local roof slaters. In the 1451 ordinary of the Incorporated Company of Slaters in Newcastle, members were enjoined that 'if any brother had taken a slate quarry or any place to cover with slates, none should undermine him, under penalty of 13s 4d.' An order was added, December 28th, 1460, that 'no brother should take less than 6s 8d for handling a rood of slate covering' (*op cit*, 696).

In the rural uplands, roofing has in the main been obtained from the Millstone Grit but, because of the small and scattered nature of the workings, it is not possible to give detailed descriptions of the geology in terms of specific locations or geological horizons as has been done for other regions. However, in the Millstone Grit the sources of stone slates were specifically identified in early texts and usually named slate sills. Forster for example, places three below the Fell Top Limestone (a bed used for correlation in this series) on a section from Newcastle-upon-Tyne to Cross Fell in Cumbria.

Figure 112. Carboniferous stone slates at Allenheads, Northumberland. The operating delph near here is in the Millstone Grit but the red colour of the covered parts indicates that the source might be in the Coal Measures (© Terry Hughes).

Figure 113. At Alston in Cumbria a small delph, Flinty Fell, produces Carboniferous slates for local use (© Terry Hughes).

> Slate sills. These strata are of a siliceous nature, will strike fire with steel, and frequently contain small particles of mica. The common grey roofing slates are obtained from the slate sills. (Forster 1883, 54)

Forster's slate sills are probably the source of stone slates at Allenheads (Fig 112) and Alston, in Cumbria (Fig 113), which are still being worked.

In County Durham one well-known quarry, Shipley Bank, near Barnard Castle, has been working the Millstone Grit for stone slates and other products for 260 years in beds about 150ft above the Upper Felltop Limestone (Mills and Hull 1976, 214).

Market

The market for stone slates in this region is not large because these areas are very sparsely populated. It is mainly within conservation areas. Demand for slates is therefore small and, regrettably and for the time being, is largely being satisfied by use of reclaimed material. Local opinion identifies the higher cost of new stone slates coupled with 'inadequate' grants which do not offset the extra cost as the overriding cause of this. It is acknowledged locally that an increase in the level of grants to fully cover the extra cost would enable tighter conservation controls to be enforced, leading to a reduction of lost roofs and an increase in roofs saved with new slates. This, of course, is exactly the strategy which has been adopted so successfully in the Yorkshire Dales Barns and Walls Scheme.

Figure 114. Carboniferous, probably Millstone Grit, at Hartburn, Northumberland (© Terry Hughes).

Historically, large parts of Northumberland, including the lowlands, were roofed with Carboniferous stone slates. Today almost all the demand is from the uplands south of Hexham in the rural areas such as Teesdale. North of here, village after village show only the remnants of a major roofing type. Often only the church has its original covering and commonly it is only the associated boiler house which is still stone-roofed (Fig 114).

The Pennine area is a Ministry of Agriculture, Fisheries and Food Environmentally Sensitive Area where traditional farming and crafts are supported but it is obvious that many roofs of farms and outbuildings are being lost to a variety of other roofing materials, mainly profiled-sheet. This is probably the source of much of the reclaimed slates which are supplying other roofs throughout the county.

There is some evidence that stone slates have been taken from roofs in this area to supply roofs in the Yorkshire Dales National Park. This is contrary to the Park's policy, but is very difficult to control if the use of reclaimed slates is permitted at all.

Production

There are four operating delves within this geographical area, although one, Flinty Fell, is in Cumbria. Only two, Shipley Banks and Ladycross, can supply with regularity and in reasonable quantities. The others, at Allenheads and Alston are small and the production of roofing is irregular.

Shipley Banks Quarry at Lartington near Barnard Castle has been working since 1862, partly underground. Currently it is untopping the old mines, with reasonable supplies of roofing available, although not always for immediate supply. It is working the Millstone Grit. [41] Some other quarries in the Barnard Castle area, including

Figure 115. Ladycross quarry south of Hexham is one of the few quarries in the north-east which can reliably produce stone slates (© Terry Hughes).

Windy Hill on the outskirts of the town, produce roofing occasionally but mainly as a self supply to building companies.

Ladycross Quarry (Fig 115) has operated for more than 300 years in the Coal Measures but was expected to run out of rock by the end of 2000. The reserve is dish-shaped at the top of a hill and therefore runs out in all directions. Below is sand and above is some coal. It has been reinstated as a nature reserve over the past 25 years and, when exhausted, is to be used as a conservation educational resource and an example of quarry reinstatement. Happily, the closure has been postponed but, if it should happen in the future, it will seriously deplete the supply of stone slates in the region and, if an alternative site should need to be found to replace it, action needs to be taken soon. There may be other suitable rock in the surrounding fells and there are five disused delves at NY 952562 and NY 952565 in Slaley Forest, at Slatequarry Clough NY 961545 and at NY 950543 and NY 947538 and several others towards Blanchlands. Delves producing slates in the Durham and Northumberland region are listed in Annex A.

Other sources

The other main source of stone slates in the region is second-hand. Even in the upland areas there are many old buildings whose roofs have been replaced with other materials to supply this trade. In 1998 there was a small importation of new and reclaimed stone slates from other parts of England, mainly Yorkshire, but there appeared to be no imports from outside the UK. Experience in Yorkshire, where imports of Indian stones have become a significant part of the market in the last few years, indicates that the situation will soon be repeated in Northumberland and County Durham.

Summary

The market for stone slates in the region is quite small and to a large extent is being supplied by cannibalizing other roofs even though it is possible to obtain authentic new slates from within and outside the region without great difficulty. Since the demand is modest and the supply comparatively satisfactory, the nub of the problem of stopping cannibalization and increasing the conservation of roofs is the lack of an adequate level of grant support to offset the extra cost of new stone slates.

CUMBRIA

Cumbria has two types of stone slates: the Carboniferous sandstones which are found on the Pennine Moors to the north of the Cumbrian Mountains, and in the south of the county around Sedburgh and Kirby Lonsdale; and the Permian sandstones of the Eden Valley. The former are the same stones as described in Northumberland and Yorkshire. The latter, the St Bees and Penrith sandstones, have supplied a small area roughly from Penrith to Carlisle, bounded to the east by the Pennine escarpment and extending some distance to the west where they were supplanted by the Skiddaw slates and the Borrowdale Tuff slates.

Geology of the Pennine uplands and north- and south-east Cumbria

In the east and south of the county and on the Pennine uplands the Carboniferous sandstones have been extensively used for roofing. These are described in the Yorkshire and Northumberland sections. However, it should be noted that one small delph, Flinty Fell, listed in that section and which produces stone slate, is actually in Cumbria, at Alston. There are similar roofs in the north at Caldbeck near Wigton (G Emerton, personal communication).

Geology of the Eden Valley

The full sequence of the Permian B Triassic in the region includes two sources of fissile sandstone: the St Bees and the Penrith. A similar stone is worked at Llocharbriggs, in Dumfries (Table 27).

> Penrith Sandstone. The quarries which abound on the rough heathland between Blaze Fell and Penrith Beacon yielded freestone, tiling, paving and rough walling stone. (Asherton and Wade 1981, 129)

There are a number of delves in the Penrith Sandstone in the Armathwaite-Lazonby-Plumpton area and in the St Bees Sandstone further east. Colloquial evidence suggests that the most durable stone slates were restricted to those between the A6 road and the River Eden at Blaze and Lazonby Fells in the Penrith Sandstone (W Clementson, personal communication). It is likely that if any roofing came out of the St Bees sandstone it was in minor

Table 27 Eden Valley sandstones

St Bees Sandstone	300+ feet (91.44 m)of bright brick-red, water-lain sandstone
Eden Shales	75–80 feet (22.8–24.3 m)of dull red mudstone and siltstone
Penrith Sandstone	0–460 feet (0–140 m)of dull brick-red, medium to coarse-grained aolian sandstone
Bockram	ill-sorted continental breccias

Figure 116. Permian sandstone at Fairbanks Farm near Plumpton, Cumbria (© Terry Hughes).

quantities. At Dufton quarry near Appleby, for example, there is so little thinly fissile rock visible today that the delph is economically negligible (David Jefferson, personal communication).

Further south, to the east of Penrith, there are several delves, one in Slatequarry Wood, which may have supplied roofing during the development of Penrith around 1800, when the local quarrying industry expanded.

The St Bees stone was formed in water environments, in a similar way to other roofing stones in England and is a durable masonry material. In contrast, the Penrith sandstone was formed from desert dunes and is very variable in quality, depending on the extent of secondary silicification.

Market

There is a moderately-sized market for Carboniferous stone slates along the Pennine Uplands and the Yorkshire/Lancashire border. A number of different specific stones were used originally but today because of the paucity of supply there is no opportunity to use authentic

Figure 118. An old Permian sandstone roof near Penrith destined to be stripped and used elsewhere. The process will inevitably lead to the complete disappearance of this distinctive roof type unless a source of new slates can be found (© Terry Hughes).

stone for reroofing in the majority of cases, unless they are taken from other roofs.

The Permian stone slates are locally significant in the villages between the M6 and the Pennine escarpment. Armathwaite and Lazonby villages are particularly rich in these distinctive red roofs (Fig 116). At Fairbank Farm near Plumpton, there is a range of buildings with the original roofs almost completely intact; a rarity in the region (Fig 117 and see Fig 22).

Elsewhere, there is evidence that these stone slates are being lost at a significant rate (Fig 118). In Penrith, old paintings show many stone roofs but, today, there are few left and most of these only have stone courses at the eaves. Interestingly, there are many old roofs covered with 'red', metamorphic slates from the Cambrian belt in Wales, presumably an early attempt to conserve the colour of the sandstone roofs as they were renovated.

Outside the area bounded by the M6 to the west, the Pennine escarpment to the east and beyond Carlisle and Penrith, stone slates become a minor feature. Most roofs are now partly covered with substitutes, mainly local metamorphic slates, the remaining sandstone having been concentrated at the eaves or on fewer slopes.

This market is being entirely supplied by reclamation from other buildings. This is bad enough for the future of this unique material, but the total number of roofs is so

Figure 117. This farm near Plumpton has one of the few remaining suites of the local red sandstone roofs (© Terry Hughes).

small that it cannot sustain cannibalization for long before it will have entirely disappeared. At present, there appear to be no initiatives in hand to establish production of new slates.

A similar stone has been used for diamond-pattern roofing in Dumfries. These also are unobtainable and the few remaining roofs will eventually be lost unless the supply problem is addressed urgently. A single quarry could supply both roofing types.

Production

Carboniferous stone slates are produced in small quantities at Alston. Two companies operate quarries in the Permian sandstone and, although they are believed to have had fissile rock at a high level in the past, neither can produce split slates at present. In the Eden Valley generally, masonry is used from the two local quarries and from Llocharbriggs. Cumbria Stone Quarries can supply roofing sawn to thickness with a flame-textured surface and dressed edges, but they report little demand for it. It is not known how long a life it would have in a roofing application. Quarry details are listed in Annex A.

Other sources

For Permian stone slates there are only two alternatives: sawn or reclaimed. New and reclaimed Carboniferous sandstones are sourced from surrounding counties. At present foreign stone slates do not appear to be in use in the region.

Summary

Eden Vale sandstone slates are distinctive and important, and are under severe threat of disappearing entirely. The market is too small to sustain a dedicated roofing delph, although it might be viable if other low-tech products, flagging, walling and rockery stone, could also be manufactured. A better option might be to negotiate a suitable specification with the existing quarry companies. This should include the Dumfries style of roofing which is in an even more vulnerable situation.

Urgent action should be taken to find a source of new stone slates to feed into the repair cycle. If this can only be achieved by using sawn material, it will be preferable to losing it completely, provided, of course, that it is durable.

The small but steady demand for Carboniferous roofs is largely supplied by reclamation.

THE CONSTRUCTION AND STYLE OF STONE SLATE ROOFS

The way stone slate roofs are constructed has evolved in response to the properties of the slates, their size, shape, thickness and roughness, and the weather which they have had to withstand. Size is important because, for small slates, the laps are more critical and the shoulders are more at risk of being close to the perpendicular joints. Conversely, larger slates will have bigger sidelaps and therefore can be used on lower pitches. Flatness is important because it is more difficult to get uneven slates to sit neatly with each other. What we see in old roofs today is the successful outcome of long trial and error and, in conserving historic roofs or building new ones, we ignore it at our peril.

The two basic forms of stone slates: large, flat, rectangular sandstones, most prominently those of the Pennines, the Welsh Marches and Horsham, and the smaller, uneven, limestones with their irregularly shaped heads, have produced two styles of roof. Broadly, the large sandstone roofs tend to have shallower pitches and simpler planforms than limestone ones, which are steeper with more complicated layouts including many raking intersections at dormers, valleys and hips. Small slates can also be made to fit around curves which are impossible for large slates. This has influenced valley design.

The best defence a slate roof has against the elements is a steep pitch. The reason for this is that the lower the pitch the more critical is the size of the headlaps and sidelaps. It is the dimension of the laps, the distance that rain or snow has to travel before getting underneath the slates, which keeps the roof dry. Because stone slates are made with random widths and diminishing lengths, and because many are irregular in shape, more so with the limestone slates, less so with most sandstones, positioning them with their neighbours on a roof and the sidelap achieved is a matter of careful judgement by the slater. This is especially critical if the theoretical laps are reduced because the top edge of the slate is shouldered or ragged. If the shoulder is too close to the perpendicular joint above, the actual laps will be too small and the roof will leak. But, if the pitch is steep there is a larger margin of safety in applying this judgement. So it is dangerous to use lower pitches than are traditional for any particular stone slate and locality.

In the limestone slate regions it is common practice to improve the cover at the heads of roughly shaped or shouldered slates by bedding small pieces of stone in mortar over any gaps. These are called shadows, shivers, shales or gallets (Figs 119 and 120).

The spread of water between stone slates has little to do with capillary action, which is more a feature of smooth-surfaced metamorphic slate roofs. It is far more dependent on penetration by wind-driven rain. The consequence of this is that stone slate roofs need to be 'wind-tight'. Traditionally, there were two methods of achieving this: using lime mortar or moss. Lime mortar was used to seal the joints between the slates inside the roof by bedding the heads (but only the heads) of the slates as they were laid and / or torching the undersides of the finished slating with lime mortar from below. Alternatively, the joints between the slates were plugged with moss, usually sphagnum or a similar species. This needed to be renewed at regular intervals and there was a recognised trade of mosser with special tools to drive in the moss. Today, roofing felt is used to make a roof wind-tight.

The oldest method of fixing stone slates is by peg hanging over split laths (Fig 121) Different woods have been used for the laths, including fir, oak and sweet chestnut. Split laths are not straight and the slight undu-

Figure 119. The use of shadows or shales bedded in mortar to improve the laps of roughly shaped limestone slates. The joints which are being protected by three shales are shown in Figure 120. The other shales have slipped from further up the roof (© Terry Hughes).

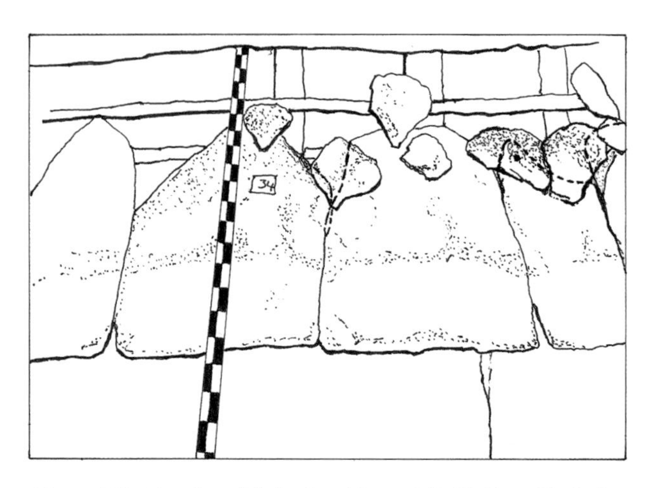

Figure 120. Drawing of shales from Figure 119 (© Terry Hughes).

lations which this produces along the slate courses are one of the attractive features of stone roofs. Pegs were most commonly oak although a variety of materials have been used including, metal and bone (Fig 122). The pegs are forced into a hole at the top of the slate. Wooden pegs should not be green because green wood will shrink as it dries and become loose. Normally they are dried before use so that they swell and become a tight fit. Pegged roofs were usually torched with lime mortar and, besides

Figure 121. This pegged roof has not been torched, consequently all the pegs have tilted, allowing the slates to slip downslope, reducing the headlaps (© Terry Hughes).

making the roof wind-proof, this also served to hold the pegs in place preventing the slates slipping. Eventually, perhaps after about 100 years, torching will become loose and fall away allowing the slates to slip. This is often the cause of leaks in a stone roof. The need to renew fallen torching should always be included in a building's inspection and maintenance programme.

From the nineteenth century sawn softwood battens began to replace split laths. Because these were thicker is was possible to nail into them and this method of fixing gradually became common. The decision to re-roof using either peg hanging or nailing depends on the conservation and technical priorities and the implications should be carefully reviewed for each roof. Nailing into battens is a more secure fixing method than peg hanging and provided non-corroding, preferably copper, nails are used the roof will have a very long life. On the other hand pegged roofs are more easily repaired because adjacent slates can be pushed aside or removed. For re-roofing, nailing also avoids the need for torching to hold the pegs which is often impossible where, for example, a ceiling which would not have been in place during the original roofing now prevents access to the underside of the roof. There are other methods of securing the pegs, such as double battening which traps the peg, but these are not

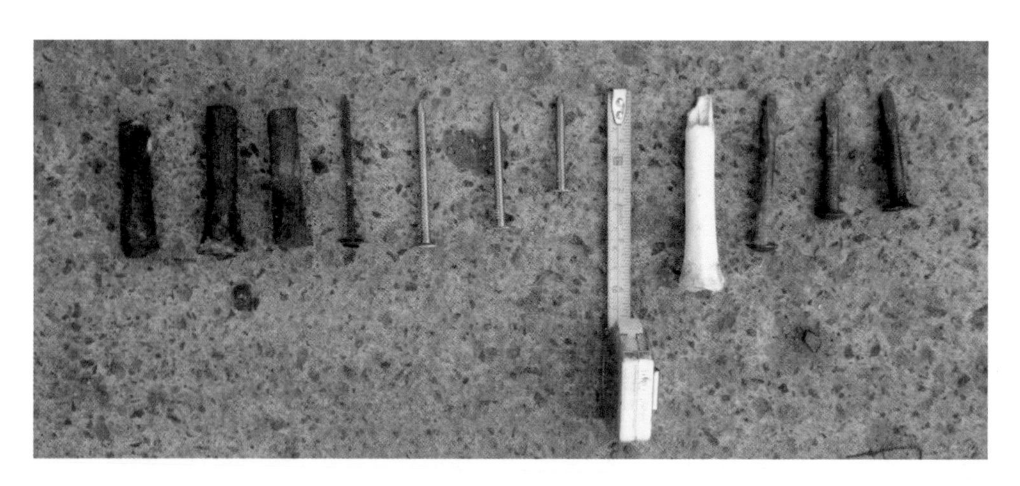

Figure 122. Some alternatives to wooden peg slates including bronze ship's rivets, a galvanised nail and washer, copper nails and bone (100 mm long), collected by Mathew Charlton, slaters of Hexham (©Terry Hughes).

Figure 123. Stone ridges were either sawn or, as in this case, adzed out of the solid (© Terry Hughes).

Figure 124. Stone hips are usually mitred and pointed with mortar but the mortar line should not be thick and obtrusive (© Terry Hughes).

suitable for small slates where the battens are too close together to fit a second one.

If pegging is chosen for reslating and metal pegs are acceptable, they can be made secure by drilling holes which are only slightly oversized or, if the original holes are larger than the peg's shank, by using pegs or nails with large heads (or a large washer) which will lock against the slate. If pegging is used over slater's felt there must be a gap between the end of the peg and the felt to avoid puncturing it. This may mean counter battens have to be introduced which raise the slating, often leading to problems at abutments, especially under copings.

Having ensured that the main areas of a stone slate roof had adequate head and sidelaps and were wind-proof, the slater could turn his attention to the careful construction of the roof intersections; the ridges, hips, abutments and, most importantly, the valleys. Because water flows away from ridges and hips but towards valleys and abutments (down the abutting wall) the former are less vulnerable and require less caution than the latter. It was the way that slaters dealt with these intersections that produced the most distinctive regional individuality. None of the traditional methods which developed relied on lead to make them work. Sometimes reliance was placed on lime mortar to form a 'seal' or to cast water away from a junction and it would be necessary to carry out periodic maintenance such as applying a lime-wash. This could produce a very distinctive effect noted by contemporary travelers with the ridge and abutments picked out in vivid white. [42]

The traditional finish for ridges is stone. These were formed from freestones, occasionally obtained from the same quarry as the slates, by adzing or sawing a V shape out of a block of stone (Fig 123). Sawing allowed several 'nesting' pieces to be cut out of one block. The ridge is mortar-bedded onto the slates, preferably without showing any mortar along their length. As the clay tile and brick industries and transport systems developed, clay ridges came to supplant stone in some areas.

Hips are usually formed with wider slates close-mitred along the hip line. The hip-joint is traditionally weathered with lime-mortar and this should be thin and unobtrusive unlike those in Figure 124. It is normal today to include lead soakers but they are usually pointed with mortar to reproduce the traditional appearance. Covered hips are quite rare and tend to be found on designed rather than vernacular buildings.

Abutments traditionally relied on a mortar fillet laid up onto the abutting wall above the slating. At side abutments, the slating was often tilted towards the roof by raising the last rafter thus directing water away from the joint. Additional protection was provided by setting slabs of stone into the abutting wall or chimney stack which helped to throw water onto the roof slope. Known as crowsteps, they also acted to some extent as a dampcourse in the wall (see Fig 103). Today abutments are normally formed with lead soakers but they look better covered with a mortar fillet rather than a lead cover flashing, especially where this involves cutting chases into the wall. To reduce shrinkage and cracking of the mortar fillet, stainless steel mesh can be incorporated.

The design and construction of valleys is critical especially if, as was normal, they are constructed without lead soakers. Several methods have evolved including swept, predominantly in the Cotswolds and the south-west; laced, on Collyweston roofs and chevron, largely restricted to the Pennines. Close-mitred and open lead valleys are not traditional and should not be used to replace any of the other types.

All the valley types have to cope with the problem of forming a mitre-joint between two slopes which involves cutting away a large amount of the slates but also providing a water-tight junction just where the most water is being carried. Although it was not always done traditionally, it is preferable to use valley boards of adequate width to support the slates across the junction.

The chevron valley, which is best suited to large thick sandstones, is formed with a row of single slates set into the slating below the mitre-cut slates in the adjacent course (Figs 125 and 126). The cut slates should be wide enough to provide an adequate sidelap with the courses above and below.

Today laced valleys (Figs 127 and 128) are similar to chevron in that they have a single wide (approximately square) slate, laid up the valley line against which the adjacent slates are set, but, unlike a chevron, they are

Figure 125. The first course of a chevron valley. The first valley slate must be long enough to extend over the gutter (© Terry Hughes).

Figure 126. In subsequent courses the slates are cut to fit against the valley slate and are wide enough to maintain adequate sidelaps with the main slating (© Terry Hughes).

butted up in the same course. The valley slate is turned through 45° so that it fits against the adjacent slates. This is an evolved form from the true lacing technique which laced slates up and over each other from left and right in successive courses. Because the valley slates were selected from the stock of whatever was available the valley line tended to waver. In the more modern form square slates are selected and centered in the valley with the adjacent slates dressed to fit to them. In a true laced valley the courses can be seen to turn upwards into the line of the valley and the adjacent slates are taper cut to accommodate this turn (see Fig 47). In a further variation on the laced theme the adjacent courses do not turn up in this way giving an appearance more akin to the chevron type. [43]

In the Cotswolds swept valleys are the normal, traditional detail (Figs 129–133). They are also used in Dorset, Somerset and with Harnage slates in Shropshire. Because of their size and shape, both pendles and presents can readily be laid into curves and this has led to the development of this distinctive valley type. In the traditional form the valley is made up with alternating courses of two and

Figure 127 (left) Side slates and valley slate. It can be seen how the side slates are turned to fit against the valley slate. Figure 128 (right) Slates are being prepared for the construction of a laced valley. Each side slate is cut to fit the central slate and to bond with the adjacent slates. (© Terry Hughes).

Figures 129. A Forest Marble roof near Cirencester showing a swept valley (© Terry Hughes).

three slates cut to a V shape. In the three slate course the centre slate is the 'long bottom' or 'bottomer', with a 'side-skew' or 'lie-bye' on either side. On the two slate course, both are known as 'broad skews' or simply 'skews'. The valley slates are longer than on the adjacent courses because the valley is longer than the rafters, but the head lap is also larger than on the main slopes to overcome the narrowness of the cut slates: typically 5 or 6 inches (125 or 150 mm) where the main headlap is 3 inches (75 mm). In effect this often created a triple lap. Traditionally, they were usually formed without valley boards with the slates supported on stone pieces laid across the rafters and bedded with lime mortar.

John Picken, with the help of William Berry, a Cotswold slater, has formalized the relationship between the pitch of the roof and the angles that the valley slates

Figures 130 (left) and 131 (right). Two courses in a Cotswold swept valley. In this example there is no valley board and the valley slates are supported on small pieces of stone (© Terry Hughes).

Figure 132.

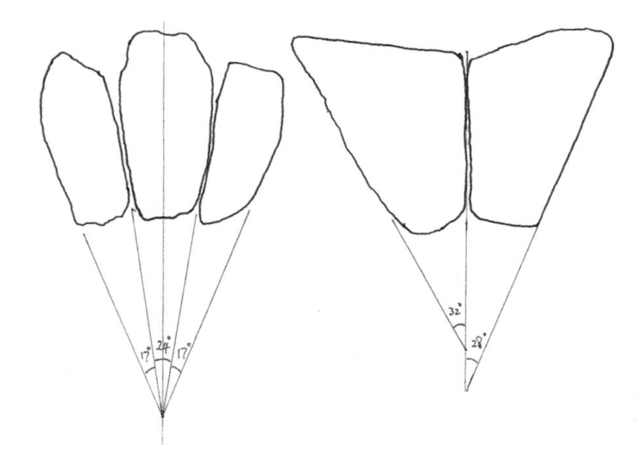

Figures 132 and 133. Swept valley slates from a 50 degree pitch roof and their angles (© Terry Hughes).

Figure 134. The ultimate colour of any stone slate will depend on what can grow on it. These lichens on a Collyweston slate are typical of calcareous slates (© Terry Hughes).

Table 28. Slate cutting in swept valleys

Roof pitch	Broad skews	Side skews	Bottomer
60°	16	8	12
59°	17	9	12
58°	18	9	13
57°	19	10	14
56°	21	11	15
55°	22	11	16
54°	23	12	17
53°	24	13	18
50°	28	15	21
49°	30	16	22
48°	31	17	24
47°	32	18	25
46°	34	19	26
45°	35	19	27

are cut to (Table 28) (John Picken, personal communication). The angles shown assume equal pitches in the two adjacent slopes and take account of the lower pitch in a valley compared to the main slopes. However, they do not allow for the fact that the slates themselves will lie at a slightly lower pitch than the valley because of their thickness and the larger head lap. So, in practice, a slightly larger angle, in proportion to their slate's thickness, would be used. Of course, being hand-cut these slates only approximate to these angles. Figures 132 and 133 show the slates from a valley on the 50° pitch roof of Lady Margaret Hungerford Almshouses, Corsham, which have angles of 17°, 24°, 17°, 32° and 28°, working from top left to bottom right.

Appearance

Stone slates vary in colour but, irrespective of this, the colour of the established roof will primarily be the consequence of the plants, mainly mosses and lichens (Fig 134), which grow on it and this is determined by the slate's mineralogy. It is also influenced by orientation to the sun, altitude, proximity to the coast and, to a lessening extent, industrial pollution. A variety of yellow, white, grey and black lichens tend to grow on limestones whereas sandstones generally look black with green algae on the north-facing slopes. However if there is mortar in a sandstone roof or if water runs onto it off the mortar on abutting walls, the alkalis it contains will promote 'limestone' lichens. Also, many sandstones are calcareous, that is alkaline, themselves and will therefore grow limestone lichens. Cement mortar promotes the growth of a thick, black moss. Where the tails and joints of stone roofs are pointed with cement mortar the resulting black moss is very disfiguring (see Fig 124). However, moss will grow in the joints of all roofs eventually, especially on shaded slopes, and it is good practice to clear this away periodically to prevent damming of the flow of water down the slates. Plants also establish and grow more readily on rough slates. The outstanding example of this is the very rough Harnage stone. Without regular attention, these roofs will quite quickly develop a dense cover of lichens, moss, ferns and even small trees (Fig 135).

The thickness and roughness of the stone slate will influence the appearance of a roof and this is accentuated by the way in which the edges are dressed. In sandstones the tails may be squared off or beveled with the beveled edge facing up or, more rarely, down. The sides can be square or beveled but the bevels may face up or down or be opposite handed on each side so that adjacent slates nest together. Limestones also show many of these variations, plus, in the Cotswolds especially, they may be beveled from both faces. This can help them to sit together more neatly. How the edges were dressed historically would have been purely the slate maker's preference, there is little technical advantage in any style except perhaps for the double beveled Cotswold slates. What is unacceptable is sawn, square edges. These are completely alien and can result in very ugly roofs. The local style should always be conserved and the detail included in the specification for any new slates.

The greatest influence on a stone roof's appearance is the random nature of the slates. All stone roofs are assembled from a mixture of lengths and widths. The mix is a natural outcome of the rock from which they are made and the way the delph is worked. The delph supplies sufficient slates to cover the area of the roof and it is the job of the slater to turn the random sizes into a well-ordered roof by deciding how they should be arranged in order to resist the weather. It is the variety of slate types, their sizes and the different slating techniques which evolved to use them which has produced the diversity of stone roofing between regions. However, stone slates from particular delves or groups of delves may have a basic similarity which creates the local or regional distinctiveness.

There is very seldom good reason to interfere with the natural mix of sizes produced by a quarry. Certainly, when renewing a roof, no attempt should be made to slavishly reproduce the existing margins or gauging. Not only is this wasteful of stone because bespoke sizes can only be made by cutting down larger slates, but it also puts the slater in an awkward position. It may only be possible to produce the specified margins by abusing the headlaps. This will probably lead to leaks which may be masked for a while if the roof is felted but will eventually show up and require expensive remedies.

The best way to specify a random roof is to use a local delph and to define the appropriate, *target* range of slate lengths. The widths of slates and the number of courses of each length will arise naturally. However, there may well be a difficulty with this when re-roofing. When a roof is stripped it is normal to have to dress off the softened head of many of the stone slates (Figs 136 and 137). This means that overall the slates will be shorter, and additional long slates will be needed to keep an approximation of the original appearance. If the delph is willing to supply extra large slates this will solve the problem but this is not good economics for manufacturers because they will be left with more small slates to sell elsewhere and these will have to be cheaper because of the greater labour needed to lay them. So special large slates will inevitably be more expensive. Care should also be exercised if slates are being bought from a merchant rather than directly from a delph. This is a rare occurrence at present but is on the increase in respect of imports from other countries especially of sandstones from India. It makes life simple for an importer to stock a small range of slate lengths but

Figure 135. The colour of stone slates may become completely obscured by plant growth. There are Harnage slates underneath these plants. The failure to clean the roof periodically has resulted in it becoming saturated and it is now collapsing (© Terry Hughes).

Figure 136. Softening and delamination at the top edge of 150-year-old Carboniferous slates (© Terry Hughes).

Figure 137. Dressing off the soft top edge of an old stone slate with a brick hammer. The slate is supported on a steel sheet which acts as a break-iron. In this instance the hammer is striking perpendicular to the surface of the stone. In other traditions the 'hammer' strikes the edges along the length of the slate (Collyweston and Herefordshire, for example) or straight onto the edges (Cotswolds) (© Terry Hughes).

Figure 138. The dark moss growing on the cement pointing of this roof in Corfe Castle gives an artificial appearance to the whole roof. Bedding and pointing a correctly constructed stone slate roof in this way is clumsy and unnecessary but some argue that it has been done for so long that it is now traditional (© Terry Hughes).

this will not produce an authentic local roof. Already very bland roofs are appearing in Yorkshire with only two or three different margins over the whole roof slope.

Limestone roofs

There is considerable variation in the size and appearance of limestone slates. Purbeck and, to a lesser extent, other parts of Dorset and Somerset produce some of the largest. Those from Stalbridge in Somerset and Brandsby in Yorkshire are often particularly heavy and uneven producing distinctly rough surfaced roofs. In contrast, in the Cotswolds region and the East Midlands the impression is of a roofscape of small sizes.

Pitches are usually between 50° and 55° but with extremes of 45° to 65°. This is important for both technical and aesthetic reasons. Almost all limestone roofs will be made up with shouldered slates which reduce to narrow sizes so steep pitches are required to ensure that water does not pass the side lap. Occasionally, roofs of much lower pitches are constructed, but this should be resisted for both aesthetic and technical reasons.

DORSET AND SOMERSET ROOFS

These usually have a simple plan form to accommodate the generally larger size of the slates, up to four feet wide, from the Purbeck, Forest Marble and the Upper Lias at Ham Hill. Dormers are either gabled, hipped or monopitched (pentice) (see Fig 28). The dormer windows are often carried down through the eaves, especially on the Isle of Purbeck. Hips are quite rare but where they exist they are usually pointed with a thin fillet of mortar. In Purbeck the fillets are sometimes unnecessarily thick and clumsy (see Fig 124).

Valleys are most often mitred with lead soakers, sometimes pointed with mortar to hide the lead, but open, lead-lined forms are fairly common. The latter is not a traditional detail. Gables often have copings, usually raised above the slating and weathered with lead soakers and flashings.

For a region with a substantial stone industry, clay rather than stone ridges are surprisingly common. Where a ridge runs into the slope above, the saddles are traditionally weathered with an upturned ridge, either stone or clay tile, but in Purbeck these have often been replaced with large, ugly lead flashings which are poorly fitted. Here, lead is also used to weather abutments and the cheeks of dormers, but is often badly detailed and constructed. Traditionally, abutments would have been mortared, sometimes with pieces of slate set in. The large cover flashings laid down onto the surface of the slating to be seen now on many buildings are both unsightly and technically incorrect. The best modern detail would be a mortar flashing over lead soakers, one to each slate. On Purbeck, small dormer cheeks look best when covered with cut-to-size stone panels over soakers. These are readily available from the local quarries.

Mortar has been used traditionally both for spot bedding to help the rougher slates sit without rocking and to fill large gaps, but it should not be visible. On the Isle of Purbeck, but not elsewhere, many of the roofs have been spoiled by excessive and exposed bedding and pointing where a heavy black moss picks out the shape of the slates in an artificial and unattractive way (Fig 138). Arguments are made for and against the use of mortar in this way. In its favour it is argued that it has been going on for a long time. The case against its use is that it can lead to problems and even premature failure of the roof timbers if the mortar cracks and water gets in and cannot get out again. What is certain is that perfectly satisfactory stone roofs can be built without the heavy use of mortar as is the case throughout the Jurassic stone slate regions from Dorset to North Yorkshire. On the mainland the slating is laid either dry or bedded. Geographically the distribution of these two methods appears to be random. Outside Purbeck, bedding mortar when used is usually raked back at the tail and perpendicular joints.

Roofs outside the quarrying areas often only have stone slates at the eaves (see Fig 30). It is suggested that in Wareham, following the great fire of 1762 which destroyed the thatched roofs, large stone slates were used to span from the rafter to the outside face of the walls when they were recovered with plain tiles (Richard Spiller, personal communication). The lower courses of the plain tiles are bedded onto the slates. Elsewhere, where stone slates are only present in the lower courses, they are the remains of the original stone roofs which have been replaced higher up with other materials, usually metamorphic slate or clay tiles. The loss of stone slates in the upper courses is particularly common at Montacute where new Ham Hill slates have been unavailable for a long time (see Fig 23). Occasionally, stone slates form the eaves to thatch. It is not clear whether this is original and/or traditional (Fig 139). In Derbyshire, Farey recorded this apparently as a deliberate technique.

> In the Northern parts of the County it is common, when Thatch is used, to lay it on a course of strong Eaves-slates or Tile-stones which prevent the Cattle from pulling the Thatch off low Buildings surrounding a yard or against a field; and ladders, &c occasionally placed against such Buildings, do less damage than to thatch Eaves; and about

Figure 139. Stone slates used to form the eaves to a thatch roof at Lillington. It is unclear whether this is an original technique or simply the rump of a stone slate roof. It has been suggested that stone eaves on thatched cottages prevented livestock from pulling down the thatch (© Terry Hughes).

Figure 140. In the Cotswolds, dormers are usually gabled or hipped (© Terry Hughes).

> Mansfield, Notts, such Eaves-slates are used to Tiled Buildings with the same view. (Farey 1811, **2** 14) [44]
>
> At Mansfield (Chesterfield) Quarry, Notts, a sort of Eaves-Slates, 20 inches to 24 inches high, are prepared for Tiled Buildings. (Farey 1811, **2** 431)

There are other references to eaveslates or eavestones. In Thorold Rogers, they are associated with eaving bricks which suggest that, in this instance at least, they were not used on the roof but as a soffit.

> Where slates were used for roofing, tiles were often employed for crests and gutters. But sometimes the stone was hewn into crests and evestones. Thus evestones are bought in Oxford at 4s. 8d. the hundred in 1410 at 5s. 9d. in 1421; at 4s. 3d. in 1435; at Radcliffe in 1419 at 3s. 4d.; in Coleshull at 5s. in 1441; and again at Oxford in 1507 at 3s. 4d.; in 1509 and 1514 at 3s.; in 1510 at 5s.; in 1518 at 4s. Enyslate or evyslate for the same purpose is bought for Radcliffe in 1437 at 7s.; in 1457 at 4s.; at Beading in 1481 at 4s. the load. In 1574 a load of eaving slate is purchased for 8s. 6d., and in 1570 a hundred of eaving bricks at 2s. 6d.; common bricks at the same place and time being purchased for 18s 6d. the thousand. (Thorold Rogers 1882, **4** 432–67)

Cotswolds roofs

There is considerable variety in the appearance of Cotswold roofs because of the wide range of slate sources. While large sizes, up to 50 inches long (1.27 m), do occur, the overall impression is of small slates when compared with other regions. Typically they range from 24 to 6 inches long (610–150 mm).

In this region slate thickness and surface roughness also have a considerable influence on a roof's appearance, although these aspects are less localized than might be anticipated. This is especially the case on more recent roofs, both new and reconstructed, where the slates may have been supplied from more remote delves than traditionally. Presents are more uneven and heavier than pendle, producing a more highly textured roof

A typical roof is gabled, sometimes with copings, but hips are not uncommon. The copings, often with elaborate kneelers and finials, are usually raised slightly above the slating with the abutment weathered with lead and mortar. Lloyd (1929, 96) has described the use of rafters steeper than the line of the copings to compensate for the reducing thickness of the slates between eaves and ridge. This ensures that the slating and the copings are parallel. Hips are normally mitred with a mortar fillet which should be thin. The traditional ridge is stone, cut from a freestone, but these are frequently replaced nowadays with clay or concrete ridge tiles. The eaves normally have a slight bellcast or cussome, which is a consequence of the rafters being footed towards the inside of the wall. There are instances of small slates being laid to fit, and emphasise the curve of the bellcast with the normal, large eaves slates commencing immediately above. This is technically dangerous as the small slates will have to carry the maximum amount of water. [45]

Dormers are most often gabled, although hips are locally common (Fig 140). As elsewhere, the junction between the dormer ridge and the main slope was traditionally weathered with an upturned stone ridge tile. This detail should always be retained with a discrete lead soaker for added security. Less commonly, the ridge of the dormer and the slates in the valley are curved up into the main slope forming a neater and more effective junction.

Although swept valleys are the traditional and most frequent detail (see Figs 127-9), laced valleys are also seen, possibly as a result of using Collyweston Slates (and slaters) to renew Cotswold slate roofs. Open lead and mitred valleys (with lead soakers) are also increasingly common in the region today. This is a recent, and undesirable, innovation stimulated by the availability of lead and the loss of the necessary skills to construct a watertight swept valley.

East Midland roofs

It is usually assumed that all stone slate roofs in the region are Collyweston. However, even though this is the most prevalent type, it is clear from the range of production

Figure 141. A typical Collyweston roofscape: hipped and gabled dormers with laced valleys (© Terry Hughes).

Figure 142. As in many parts of England where stone slates have been unavailable for a long time, other roof slates or tiles gradually replace the stone slates during successive re-roofings. This example near Kilburn in North Yorkshire shows clay pantiles over Brandsby slates (© Terry Hughes).

Figure 143. Pentice dormers in Coxwold, North Yorkshire (© Terry Hughes).

sites and their geology, that a variety of other slate types have been used in the past. Unfortunately, some types such as Duston, seem to have completely disappeared. Ultimately, of course, the appearance of the different roofs will depend to a large extent on the vegetation which the mineralogy supports. Since they are all calcareous, this is likely to be very similar for all types.

Roofs are normally steeper than 45° and gabled rather than hipped (Fig 141). Gables are finished by carrying the slates over the wall and forming a mortar fillet underneath, or with copings which may be either immediately on the slates or raised above and the slates butted up with a fall towards the roof slope. The abutment is usually weathered with lead soakers and mortar flashing. Dormers are common and often carried below the eaves (see Fig 141). They may be gabled or hipped. Dormer ridges are often swept up into the slope above. Ridges are formed with stone or clay tiles. Valleys are laced or occasionally swept when traditionally executed.

Collyweston roofs are often formed with mortar. A slate recovered from an archaeological site dating from not later than the fifteenth century was found to have been mortar bedded and laid at one third lap [46] (Williams 1979). The mortar is used to bed the slates (spot bedding) and, where there is a gap underneath the slates, especially at the eaves, shales are inserted in the bedding to support the slates and prevent breakage. It is also common practice to tail bed slates and mortar the perpendicular joints. This technique appears to be a long-standing and successful practice even if it is not strictly technically sound.

North Yorkshire roofs

In the Brandsby region of North Yorkshire most of the remaining roofs are 'designed' rather than vernacular. They include an almshouse, Newburgh Priory, churches, estate and large farm houses. It is probable that all the remaining roofs have been renovated and so the original constructional details may have been lost. Many roofs have stone slates at the eaves only, the remainder usually being covered with clay pantiles (Fig 142). Commonly, the Brandsby slates have been replaced with Carboniferous sandstones.

Generally, the design is simple, gable to gable, but some more elaborate shapes exist. Some of the latter are recent. Gables are plain or covered and the rare examples of hips are covered with tiles. Dormers are also rare and may be recent additions. An older example in Coxwold is monopitched (Fig 143). Today all valleys are open and lead-lined.

Sandstone roofs

In contrast to the limestones, the diversity of sandstone roofs is more to do with the characteristics of the stone slates themselves than to variations in constructional techniques. The stone slates are, on the whole, large and flat with only slight surface irregularities which have often been dressed off to improve the laying qualities. This permits their use at lower pitches than limestones, about 35° to 40° although in the north, on the high moors of Cumbria, County Durham and Northumberland they seldom drop below 45°.

There has been little need to use mortar in the laying of sandstone slates apart from torching which was to keep out the wind. The one exception is Horsham stone which is a special case. Even the excessively rough Harnage stone can be successfully laid with almost no spot bedding. Today, especially where slater's felt is incorporated, some local authorities discourage the use of mortar entirely.

Horsham roofs

Horsham stone slates are large and often have an uneven or rippled surface. They weather to a pale to dark brown colour (Munsell 10YR 7/4 to 4/6) and support grey and orange lichens. When heavily mortared, an unsightly black moss develops along the joints and tails of the slates.

Constructional details [47] are similar to other large-format stone slates although the pitch is usually steeper, often more than 45°. Most commonly, the roof is gable to gable (see Fig 55) but hips and valleys are frequent. The former are soakered and pointed. Valleys are either close mitred or open and lead-lined, although the latter are probably not original details. Dormers are rare. Abutments are finished with either lead or with mortar flashing over lead soakers.

Historically, because of difficulties in getting the slates to lay correctly, spot bedding with breeze [48] has been used. The use of shadows or gallets was also commonplace to improve the weathertightness where slates were excessively shouldered. In their earliest form, shadows were thin pieces of stone but now, in this region, they are normally metamorphic slate. The shadow is sometimes nailed to the same batten as the stone slates, sometimes to a separate thinner batten.

Following the demise of the quarrying industry and as supplies of reclaimed slates diminished, shadows, originally used to ensure watertightness, started to be used as a means of reducing the head lap to make the slates go further. In extreme cases, the construction was no longer double lap and the shadow slate was used simply to weather the perpendicular joint (Figs 144 and 145). [49] In effect, it has become a soaker. This technique is undesirable on technical grounds but it has been argued that it is now so well established in this region that it should be regarded as traditional and worthy of conservation. To further eke out supplies of stone slates some roofers use concrete imitation slates for the under eaves course.

The use of mortar has a long history for Horsham slating.

> Horsham-stone. ... Is a kind of thin broad Slate of a greyish Colour much used in some parts of Sussex formerly, not only to heal or cover Churches and Chancels but some great Houses also; it is call'd Horsham-stone in that County, because it is for the most part brought from a Town there called Horsham; this sort of Stone, or Slate rather, is laid of different Sizes, viz. From 8 or 9 Inches to 24 Inches, or more in length, or breadth etc. It is commonly from 1/2 Inch, to 1 Inch thick. ...... The Price of laying a Square and Pointing (which is striking Mortar under the lower ends) in new Work, 5 or 6s. But to rip it from old, and new lay and point it, not less than 6 or 7s per Square, which is the lowest I ever knew it done for. (T N Philomath 1763, 164)

Its use has been considerably extended in more recent times. What was originally a means of settling in the slates has become an integral part of the weathering. This has the disadvantages of looking unsightly and may hold or draw water into the roof and especially to the battens where it will promote rot. Brook reports that full bedding and pointing of the perpends is the oldest method used in the area, having been in existence for at least 100 years. If full bedding must be used in either double or single lap, it is far better to hold the mortar back from the edges of the stone slates and to rake out the perpends. This at least provides a route for water to drain out between the slates, should it get past any cracked mortar.

Figures 144 (left) and 145 (right). The 'single lap' method of roofing often adopted around Horsham. Welsh slate is being used as a 'soaker' to weather the perpendicular joints in the inadequately head-lapped stone-slating. A bed of mortar is laid to receive the next course (© Lisa Brooks).

Figure 146. Disintegration of a weak mortar mix in Horsham slating (© Lisa Brooks).

Figure 147. Harnage slate from the Hoar Edge Grit quarried at Acton Burnell. Munsell 2.5Y. Scale in centimetres (© Terry Hughes).

Figure 148. Close-up of Harnage slate showing the plane of fissility along the brachiopod shells (© Terry Hughes).

There is considerable misunderstanding and disagreement about the type of mortar which should be used with Horsham slating and both strong and weak mixes have been implicated in some failures (Fig 146). This issue needs to be resolved.

Welsh Marches roofs

The stone slates of the region are of Ordovician, Silurian, Devonian and Carboniferous age. (There are no known remaining roofs of Permian Age, Grinshill Stone.) The Ordovician slates from South Shropshire and the Pennant sandstones of South Wales and Bristol are robust, large and flat but often with a rippled surface. Their colours range from blue-grey to various shades of brown and red. The Devonian sandstones are grey-green or red-purple, fine-grained and micaceous. They are smaller and somewhat thinner than other sandstones. The Harnage stone slates from the Hoar Edge grit are distinctly different. They are extremely rough textured with brachiopod shells exposed at the surface (Figs 147 and 148). The colour is light brown with white mottling. In appearance they have more in common with the heavier limestone slates such as Forest Marble than sandstones.

In the Forest of Dean today, roofs are sometimes found to be a mixture of Pennant and Devonian sandstones. It is not known whether this is an original feature, implying that slates from both local sources were used indiscriminately.

Considering the variety of stone slate sources within the Welsh Marches, there is less variety in the roof forms than might be expected (see Figs 3, 60, 63, 65–68, 74 and 75). This, with the exception of Harnage stone slates, is probably a reflection of their similar size and flatness. The colour of the stone is often not apparent because of the growth of algae, lichens, mosses and in the case of Harnage, ferns and even tree saplings. Generally the appearance of the surface of the roofs is flat but with the dressed edges providing prominent detail. Pitches are steep, usually more than 45°, and typically gables, except for churches, are without copings. Hips are quite rare. Valleys are fairly common and are almost always formed with lead today, although this would not be the original detail. On Harnage roofs the valleys are swept. There are examples of these at Pitchford Hall, Langley Gatehouse near Acton Burnell and Wilderhope Manor near Much Wenlock. Dormers, which are generally uncommon, are gabled (see Figs 65, 68 and 74). Ridges are mainly stone but have often been replaced with clay or concrete. Examples of curved slating can be seen at St Briavels Castle [50] and Wilderhope Manor.

Pennine roofs

Although the roofs of this region are broadly similar in style there is considerable variability in detail because they encompass so many stone sources and stone slate types. Usually the slates are large, but there are exceptions especially in the Peak Park; at Abney and Eyam for example. Colours range from grey through buff to dark brown often with mottled iron oxide staining. In Northumberland there is a red stone in the Coal Measures which is thought to have produced some roofing but this has not been confirmed (Fig 112 may be an example). Surface texture is most commonly smooth and fine grained or gritty but there are many examples of rippled (see Fig 15), bioturbated [51] (see Fig 90) or just generally rough-surfaced stones.

The way the edges of the stone slates are dressed varies across the region. Examples can be found with the tails and edges square (Fig 149) or beveled and the bevels facing up, down or opposite handed and nesting. The latter are prominent in Northumberland (Fig 150).

Roof pitches are 35° or 40° in the south of the region but about 45° in the northern counties. Buildings are simple on plan. Ford (1919 and 1925) suggested that the difficulty of making a satisfactory and weathertight joint at the intersection of a ridge with the slope of a higher roof, led to the practice of keeping the ridges at the same level and varying the pitches of the wide spans of the rooms below, often producing in the same house gables of widely differing slopes. Dormers are rare and there are few valleys or hips. Where valleys do exist they are traditionally of chevron type (Figs 125 and 126) but today they are often replaced with lead. Hips are rare, and gables are either plain or with copings. The copings are usually laid directly on the roof slope but sometimes they are raised up slightly. A common feature of the larger houses and halls of the region is for gable walls to be carried above the roof slope with a stone slate abutment under carved copings often with kneelers and elaborate finials. [52] Traditionally, such abutments are finished with a mortar flashing under the coping. The traditional stone ridge is now commonly replaced with clay tiles.

Some roofs, especially in Northumberland, have been mortar pointed in recent years to keep out the wind but as roofs are reslated this detail is being omitted in favour of slater's felt.

Figure 149. The way the edges of slates are dressed varies across the Pennines. Square-edged slates on a Derbyshire roof (© Terry Hughes).

Figure 150. Opposite-handed edges at Blanchlands in Northumberland (© Terry Hughes).

CUMBRIAN ROOFS

The two regional stone slates are distinctly different in colour. The Permian sandstone is red usually with yellow lichens and moss attached (Fig 22), whereas the Carboniferous is grey to black when weathered. Otherwise, in form and construction they are similar.

The Permian sandstone slates are large. It is not unusual to find roofs with a length range from about 40 down to 18 inches (1015–450 mm). The roof design is correspondingly simple, gable to gable. Verges are normally oversailing but copings are seen occasionally, either laid on the slates or raised and with lead soakers and flashings. Valleys are rare; hips are less so and are covered with lead or tiles. Ridges are normally stone. Beyond the immediate quarrying area, it is common to find stone slates at the eaves only, the upper courses being covered in metamorphic slates (Fig 151).

The Carboniferous sandstone roof details are the same as those described in the Pennine section They also often show pointing but this again is not an original feature.

Figure 151. In the Eden valley, because there is currently no source of new stone slates roofs they are being replaced by metamorphic slate or tiles. Unless action is taken to find a way of producing new stone slates there will eventually be no examples of this important regional roof type left (© Terry Hughes).

CONCLUSION

Stone roofs are an important element of the regional character of the built landscape but they continue to be lost and degraded everywhere because of lack of new stone slates. While the cost of hand-made products will always place a limit on how many roofs can be conserved, other factors are contributing to this decline. These include the limitations of conservation planning control and, most importantly, the willingness of many bodies to permit the cannibalization of one roof to supply another. There is no doubt that the best use for old stone slates is to put them back on the building from which they came. To continue to rely on reclaimed slates to maintain old roofs is simply not sustainable and eventually they will all be lost.

Authentic building conservation demands the use of authentic materials. For most of a building's fabric it is possible to obtain these. Even if it takes hundreds of years to grow an oak tree, it can be done. But for authentic stone to be available there must be delves. The quarrying industry has a poor reputation with the public and planning applications even for small and unobtrusive stone slate delves always elicit unthinking public objections, often from the very people who live by choice in stone-roofed, -walled and -flagged houses. Unfortunately, much of this negative attitude to delving is being promoted by environmental protection legislation. Primarily directed at large-scale aggregate and minerals operations, its rigid application to small 'conservation' delves will be to the detriment of our vernacular building heritage. To prevent the operation of a small delph in a National Park on environmental grounds and by so doing promote the importation of stone from thousands of miles away, makes no environmental sense at all.

In spite of the difficulties in producing stone slates there has been considerable success since the English Heritage *Roofs of England* campaign began in 1994. Delves are now operating for the first time in many years in the Peak Park, Herefordshire and South Wales. In the Cotswolds and the adjacent counties which use limestone slates, the supply is assured in a way that it has not been for about fifty years and for several different stone slate types. In the Cotswolds and the Pennines those active delves which predate the campaign continue to be busy supplying a healthy demand. There is also good reason to anticipate the re-establishment of commercial production at Collyweston and Horsham.

The temporary operation of a delph for Harnage stone has demonstrated that roofs with minor but important stone slates can be successfully conserved. This bodes well for other minor slates such as those from the Eden Valley and the Magnesian Limestone of eastern Derbyshire.

It can reasonably be concluded that the campaign to promote the manufacture of stone slates has now reached a point where it is time to look at the market and to tackle some of the demand-side problems. The issues of the correct, local and regional, design, specification and construction of stone slate roofs is of paramount importance here. Little formal training of any depth is available for this and there are few reliable written sources of guidance on the detail of stone roof construction. This, together with rapidly changing building control legislation and the ill-informed promotion of new roofing products and accessories, is a recipe for roof failures.

The Stone Roof Working Group is attempting to tackle some of these problems but will need help from experienced workers in the regions. Most of the twentieth century saw the decline of vernacular stone roofing and the loss of thousands of stone roofs. Because of the efforts of a few individuals and organisations, this trend has started to be reversed but, there is no doubt that the best hope for the future is a determined effort by local enthusiasts with a willingness to persevere in the face of bureaucratic, technical and financial obstacles often over many years.

ENDNOTES

1 This report covers stone slates, ie sandstones, limestones and, rarely, igneous rocks used for roofing. It does not include the true, metamorphic, slates of Cornwall, Devon, Leicestershire, Lancashire and Cumbria.

2 (Pevsner 1973) Apethorpe was probably not Collyweston *senso stricto*. They probably came from Duston (Sussex Archaeological Collection 54.152).

3 'The term Peterborough is generic: it indicates the warfage point for stone other than Barnack, and for the slate, but not necessarily the location of these materials ... Gunwade being solely for Barnack stone at this time, the slatestone was negotiated through a Peterborough factor and keeled from the Peterborough wharves, thus gaining its misleading epithet' (Sharp 1982, 64).

4 Beaurepaire is now known as Bearpark, about two miles from the centre of Durham, and is the site of the Prince Bishop of Durham's hunting lodge (David Jefferson, personal communication).

5 Stone Roofing Working Group guides are available from Terry Hughes, Slate and Stone Consultants, Ceunant, Caernarfon, Gwynedd LL55 4SA or at www.stoneroof.org.uk

6 Geological formations are often named for the locality in which they were first described. As a result Purbeck beds (and stone slates) are found on the Isle of Portland and Portland beds occur in the Isle of Purbeck.

7 Arkell and Tomkieff (1953, 106) suggested that the word shiver derives from the Middle English 'scifre' meaning a splinter, chip or fragment and hence shaley or slaty debris. They note the similarity to the German 'schiefer' meaning slate.

8 See also a comprehensive bibliography and guides to the geology of the region at www.soton.ac.uk/~imw/

9 Pennant sandstone-slate near Bristol.

10 This quote uses a more restricted geographical limit to the Cotswolds than used in this study. Good stone slates are delved outside the Cotswolds *senso stricto*.

11 Facies: the sum total of features such as sedimentary rock type, mineral content, sedimentary structures, bedding characteristics, fossil content etc which characterize a sediment as having been deposited in a given environment.

12 A square is 100 square feet of roof area (9.29 metres^2).

13 Howe (1910, 318) incorrectly describes this site as being in Wiltshire.

14 Planking: any coarse, flaggy stone that can be used for flagging (Arkell and Tomkieff 1953).

15 Racy clays: small calcareous concretionary nodules common in brick clays (Arkell and Tomkieff 1953).

16 Atford = Atworth.

17 This is probably the recently reopened and now closed Botleaze Wood quarry.

18 Howe misquotes Swindon as Lumdon in *The Geology of Building Stones* (Howe 1910, 324).

19 A more detailed description of the manufacture of pendle slates is given in Aston 1974, 41–58.

20 As far as possible the stone was kept underground until the winter to avoid having to bury it.

21 Today, Woodward's use of the name Stonesfield Slate at Througham is regarded as incorrect. The rock here is the Througham Tilestone. Locally they were known as Bisley slates.
22 This company's quarry at Botleaze near Atworth failed to produce adequate amounts of stone slates and has been closed.
23 Futher information on Dr Sutherland's report may be obtained from terry@slateroof.co.uk.
24 +44 1380 840092 for details.
25 In this context pendle means fissile, not frosting.
26 The full report is held by Guildford Borough Council. The author would be very grateful for any further information on this stone slate.
27 For a variety of lithological reasons the stone is classed as a siliceous limestone. This confirms Dunkerton's observation that Horsham roofing stone needs lime in it to 'harden it into a granity stone'. In appearance however it is regarded as a sandstone.
28 Clifton Taylor also describes the Hoar Edge slates at Stokesay as small, 'only about 8 ins long'. However, it is clear from *The Pattern of English Building* that he is referring to the margin not the length; 'about 8 ins deep with 4 ins overlap' which gives a length below the fixing hole of 20 inches (508 mm) (Clifton Taylor 1962, 142).
29 The stone-slates examined petrographically in the Pitchford Church investigations were shelly sandstones. However Toghill (1990, 82) has reported that they have sometimes been described as limestones. 'Within the Hoar Edge Grit calcareous sandstones occur in the northern area of the outcrop around Coundmoor and Harnage. These have sometimes been called the *Subquadrata* Limestone because of the abundance of the brachiopod *Orthis subquadrata* (*Harknessella subquadrata*). They split into thin slabs used for roof tiles'.
30 The report mentions Hill Top quarry specifically. Although the title of the paper refers to the Downtonian, on the basis of the fossil *Pterapsis leathensis* the roofing appears to be in the Dittonian.
31 Bettws is taken to be Bettws y Crwyn, south-west of Bishops Castle in the Přídolí.
32 The Welsh word llechau is equally translatable as slate, an option which is understandably avoided by a geologist.
33 From their third report.
34 Howe uses the name Burley Down. This appears to be a transcription error.
35 Diagenesis: changes affecting a sediment while it is at or near the surface. These include dewatering and the cementition of rock grains.
36 Care is needed in interpreting Walton's geographical references.
37 An early reference (1439) uses the name Baxstondelf, a possible corruption of thackstonedelf or bakestonedelf. Thackstones are roofing stones, bakestones are made from similarly fissile rock. 'The word thatch itself is a weakened form of thack, which, in turn, is almost pure Anglo-Saxon, and cognate with the Latin 'tegere', its prime meaning being therefore a 'covering', irrespective of the material (Ford 1924).
38 The word mine can be used to describe a surface quarry or delph.
39 The presumption that stone roofing was largely restricted to more significant buildings has arisen because of the type of documents analysed by Salzman and others.
40 For a detailed description of Scottish roofing see Emerton 2000.
41 Not the Coal Measures, as reported in Leary 1986, 64.
42 Walton (1941, 95) includes a picture of white pointing.
43 H B Sharp (1980) has described Collyweston slating, including valley construction.
44 The comment that eaves slates are used to protect tiles is probably incorrect. It is more likely that their purpose was to span to the outside face of the wall from the rafters footed onto the inside face or that they were the remnants of a former stone roof.
45 A example at Combe Down, Bath is shown in anon (nd), 10.
46 One third lap is where the lap is a third of the length of the slate. It should not be exceeded in normal double lap slating because it then becomes triple lap which, although it is necessary in special circumstances, is wasteful of slates.
47 There is a detailed description of Horsham slating and a series of photographs in Brooks 1997.
48 A mixture of crushed clinker and lime or cement.
49 Double lap slating is the system in which course three overlaps course one by the headlap dimension, four overlaps two etc. In Horsham the double lap system has become single lap, course two overlaps course one by the headlap dimension. This leaves the perpendicular joints open resulting in the need for large shadows or soakers.
50 At St Briavels Castle the tower is too tall for the curved slating to be seen from the ground.
51 Bioturbation is the disturbance of sediments by aquatic animals. This leaves traces in the rocks formed from the sediments which, when the rock is split, show as positive or negative casts. Worm casts are one example.
52 Walton 1941, 66 and 93–96 describes the detail of Pennine gables.

ANNEX A QUARRIES WORKING IN STONE SLATE ROCK

REGION	QUARRY NAME	COMPANY DETAILS	TYPE OF STONE	COMMENTS
Dorset & Somerset	Cobb's	Cobbs Quarry, 28 Eastington Road, Worth Matravers, Swanage, Dorset BH19 3LF	Jurassic - Middle Purbeck	
	Down's	Harden Bros, Downs Quarry, Kingston Road, Langton Matravers, Swanage, Dorset BH19 3JW	Jurassic - Middle Purbeck	
	Blackland's	H F Bonfield & Sons, Blacklands Quarry, Acton, Langton Matravers, Swanage, Dorset BH19 3LD	Jurassic - Middle Purbeck	
	Swanage Quarries	J Suttle Quarries, Panorama Road, Swanage, Dorset BH19 2QS	Jurassic - Middle Purbeck	
	Keate's	Keates Quarry, 31 Eastington Road, Worth Matravers, Swanage, Dorset, BH19 3LF	Jurassic - Middle Purbeck	
	Down's	D & P Lovell Quarries, Downs Quarry, Kingston Road, Langton Matravers, Swanage, Dorset BH19 3JP	Jurassic - Middle Purbeck	

	Lander's Eastington	Landers Quarries Ltd, Kingston Road, Langton Matravers, Swanage, Dorset BH19 3JP	Jurassic - Middle Purbeck	
	Ham Hill	Ham Hill Stone Co Ltd, Hamdon House, Westview Road, Sparsholt, Winchester, Hampshire SO22 5RB	Middle Lias	
	Tout	Ham and Doulting Stone Co Ltd, Tout Quarry, Charlton Adam, Somerton, Somerset TA11 7AN	Lower Lias, Hamstone	
	Stalbridge	Stalbridge Quarry Ltd, Inwood House, Henstridge, Temple Coombe, Somerset BA8 0PF	Forest Marble	
The Cotswolds	Goldhill Quarry	Cotswold Stone Tile Co Ltd, Wedhampton Manor, Devizes, Wiltshire SN10 3QE	Forest Marble	
	Grange Hill, Naunton	Grange Hill Quarry, Naunton, Cheltenham, Gloucestershire GL54 3AY	Fullers Earth Formation-Eyford Member and Taynton Stone (presents)	
	Brockhill, Swellwold and Tinker's Barn	Cotswold Stone Quarries, Brockhill Quarry, Nr Naunton, Cheltenham, Gloucestershire, GL54 3BA	Fullers Earth Formation - Eyford Member and Chipping Norton Limestone	
	Soundborough	Soundborough Quarry Ltd, Soundborough, Andoversford, Cheltenham, Gloucestershire GL54 5SD	Fullers Earth Formation-Eyford Member	
	Veizey's	Tetbury Stone Co Ltd, Veizeys Quarry, Avening Road, Tetbury, Gloucestershire GL8 8JT	Forest Marble	unable to produce roofing at the time of printing
	Huntsman's	Huntsman's Quarries Ltd, The Old School, Naunton, Cheltenham, Gloucestershire GL54 3AE	Jurassic-Fullers Earth Formation-Eyford Member, Chipping Norton Limestone	
East Midlands		D W Ellis, The Rosery, Ryhall, Stamford, Lincs PE9 4HE	Collyweston Stone Slate	
		C N Smith, Blue Bell House, Collyweston, Stamford, Lincs PE9 3PW	Collyweston Stone Slate	
Welsh Marshes and Bristol	Gwrhyd Quarry, Rhiwfawr	Gwrhyd Specialist Stone Quarry, Rhiwfawr, Swansea SA9 2SB	Carboniferous - Pennant Sandstone	
	Coed Major	Coed Major Quarry, Coed Major, Craswall, Longtown, Herefordshire HR2 0PX	Devonian - (Old Red Sandstone)	not producing at time of writing
	Grigland and Pennsylvani	Grigland and Pennsylvani Quarries New House Farm, Michaelchurch, Escley, Hereford, HR2 0PU	Devonian - Old Red Sandstone	
Derbyshire & the Peak Park	Fulwood Booth	Fulwood Booth Quarry, West Carr Cottages, David Lane,Fulwood, Sheffield S10 4PH	Carboniferous - Coal Measures	
	Bretton Moor Quarry	Bretton Moor Quarry, Sheperds Flat Farm, Eyam, Hope Valley, Derbyshire S32 5QS	Carboniferous Sandstone	
Lancashire		Ford Stone, Albert Mill Depot, New Line, Bacup, Lancashire OL13 9SA	Carboniferous - Lower Coal Measures: Dyneley Knoll Flag	
Yorkshire	Hill Top Quarry	Brogden Stone Supplies, Borren House, South Stainmore, Kirby Stephen, Cumbria CA17 4DJ	Carboniferous -Millstone Grit: Stainmore Group, Tan Hill Grit	
	Tuck Royd	Rand & Asquith, Tuck Royd Quarry, Halifax Road, Brighouse HD6 2PL	Carboniferous - Coal Measures: Elland Flags	
	Soil Hill Quarry	Soil Hill Quarry, 26 West Lane, Thornton, Bradford BD13 3HX	Carboniferous - Coal Measures: Elland Flags	
	Bolton Woods Quarry	Pickard Group, Fagley Lane, Eccleshill, Bradford, W Yorkshire BD2 3NT	Coal Measures - Gaisby Flags	
Durham and North-umberland	Shipley Quarry	Shipley Quarry, Rose Cottage, Lartington, Barnard Castle, Co Durham DL12 9BP	Carboniferous - Millstone Grit	
	Ladycross Quarry	Ladycross Stone Co, Slaley, Hexham, Northumberland NE47 0BY	Carboniferous Lower Coal Measures	
Cumbria	Flinty Fell Quarries	Alston Natural Stone, Hodgson Brothers, Nenthead, Alston, Cumbria CA9 3TJ	Carboniferous - Millstone Grit	

REFERENCES

Aikin J, 1795 *A Description of the Country from thirty to forty miles round Manchester containing its geography natural and civil; principal productions; river and canal navigations*, London, John Stockdale.

anon, 1891 Ramble in Green Clough, *Bacup Times* 1st August.

anon, undated *Cotswold Stone Tiling*, Stroud, Gloucestershire, Freeman and Sons.

Arkell W J, 1945 The names of the strata in the Purbeck and Portland stone quarries, in *Proceedings of the Dorset Natural History and Archaeological Society*, **66**, 158–168.

Arkell W J, 1947 *Oxford Stone*, London, Faber and Faber

Arkel W J, Wright C W and Melville R V, 1947 *Geology of the Country around Weymouth, Swanage, Corfe and Lullworth*, Memoir of the Geological Survey, London, HMSO.

Arkell W J and Tomkieff S I, 1953 *English Rock Terms*, Oxford, Oxford University Press.

Asherton R S and Wade A J, 1981 *Geology of the Country around Penrith*, London, HMSO.

Aston M A, 1974 *Stonesfield Slate*, Oxford, Oxford County Council Museum Services.

Bailey J and Culley G, 1805 *A General View of the Agriculture of the County of Northumberland*, Newcastle, Sol Hodgson.

Bailey J, 1810 *A General View of the Agriculture of the County of Durham*, London, Richard Phillips.

Baldwin A, 1998 *Where did the Slate Roofs of Rossendale come from? Review of Some Historical Documents in Rawtenstall Library*. Unpublished report, available from terry@slateroof.co.uk.

Boneham B F and Wyatt R J, 1993 The stratigraphical position of the Middle Jurassic (Bathonian) Stonesfield Slate of Stonesfield, Oxfordshire, in *Proceedings of the Geological Society*, **104**: 2, 123–6.

Bolton H, 1890 *The Geology of Rossendale*, Bacup, Lancashire, Tyne and Shepherd.

Bristow C R, Barton CM, Freshney EC, Wood CJ, Evans DJ, Cox BM, Ivimey-Cook HC, and Taylor RT, 1995 *Geology of the Country around Shaftsbury*, Memoir of the Geological Survey, London, HMSO.

Brooks L J, 1997 *Horsham Stone Roofs*, unpublished RICS Dip Build Cons dissertation, College of Estate Management, Reading.

Carlisle J D, 1990 *A Study of the Viability of Stone-slate Production in the Richmondshire Area, Leyburn*, 2 vols, Yorkshire Dales National Park, York.

Carver M O H, 1983 Two town houses in medieval Shrewsbury, in *Transactions of the Shropshire Archaeological Society*, **61,** 32.

Clarke R B, 1950 The geology of Garnons Hill and some observations on the formation of the Downtonian rocks of Herefordshire, in *Transactions of the Woolhope Naturalists Club*, **33**, 97–111.

Clarke R B, 1951 The geology of Dinmore Hill with a description of a new myriapod from the Dittonian rocks there, in *Transactions of the Woolhope Naturalists Club*, **34**, 222–37.

Clifton Taylor A, 1962 *The Pattern of English Building*, London, Faber.

Clifton Taylor A, 1983 *English Stone Building*, London, Gollancz.

Cope J C W, Clements R S and West I M, 1969 Guide to Dorset and South Somerset, in *International Field Symposium on the British Jurassic, England*, Geology Dept, Keele University, Keele, A63–A64.

Crane T 1979 *The Stones of Newcastle*. Dissertation in Conservation Studies, University of York Institute of Advanced Architectural Studies, York.

Damon R F, 1884 *Geology of Weymouth, Portland and the Coast of Dorsetshire from Swanage to Bridport-on-the-Sea, with Natural History and Archaeological Notes*, 2nd edn, Weymouth, R F Damon.

Davey N, 1976 *Building Stones of England & Wales*, London, Bedford Square Press.

Davis D. C., 1912, *Slate and Slate Quarrying*, London, Crosby Lockwood and Sons.

Davis W, 1815 *General View of the Agriculture and the Rural Economy of Wales*, London, McMillan.

Dickson R L and Stevenson W, 1815 *General View of the Agriculture of Lancashire*, London, McMillan.

Dobson C G, 1960 *Historical Notes on the Langley Museum*, London, Langley Tiles Company.

Dunkerton F J, 1945 Horsham Stone and Sussex Marble, in *Quarry Manager's Journal*, **Nov**, 213–14.

Earp J R and Hains B S, 1971 *British Regional Geology, The Welsh Borderland*, London, NERC Institute of Geological Sciences.

Emerton G, 2000 *The Pattern of Scottish Roofing*, Edinburgh, Historic Scotland.

English Heritage, 1998 *Stone Slate Roofing*, Technical Advice Note, London, English Heritage.

Farey J, 1811 *A General View of the Agriculture and Minerals of Derbyshire*, London, McMillan **1** and **2**.

Ford T F, 1919 *Some buildings of the 17C in the Parish of Halifax*. Unpublished MSc thesis, Royal Institute of British Architects, London.

Ford T F, 1924 Some buildings of the 17C in the parish of Halifax, in *Thoresby Society Miscellanea* **28,** 30–37.

Forster W, 1883 *A Treatise on a Section of the Strata from Newcastle-upon-Tyne to Cross Fell*, Newcastle-upon Tyne, A Reid.

Fox-Strangeways C, 1892 *Jurassic Rocks of Britain. Vol I: Yorkshire*. London, HMSO.

Fox-Strangeways C, Cameron A G and Barrow G 1886 *Geology of the Country around North Allerton and Thirsk*, London, HMSO.

Gallois R W and Worsam B C, 1993 *The Geology of the Country around Horsham*, Memoir of the Geological Survey, London, HMSO.

Godwin C G, 1984 *Mining in the Elland Flags, A Forgotten Yorkshire Industry*. British Geological Survey Report London, HMSO.

Greig D C, Wright J E, Hains B A and Mitchel G H, 1968 *Geology of the Country around Church Stretton, Craven Arms, Wenlock Edge & Brown Clee*, London, NERC Institute of Geological Science.

Hart C, 1971 *The Industrial History of Dean*, Newton Abbot, Devon, David and Charles.

Howe J A, 1910 (2001) *The Geology of Building Stones*, London, Edward Arnold, facsimile edition, Shaftesbury, Donhead Publishing.

Hugh, 1272 *The feoffment of Hugh son of Margery le Crone*. Public Record Office E326/1294.

Hughes T G and Jefferson D, 1998 *Potential Sources of a Traditional Roofing Material in the Hoar Edge Grit, Shropshire*, unpublished report for English Heritage, London.

Hughes T G, Thomas I A, Guion P D, O'Beirne A M and Watt G R, 1995 *Roofing Stones in the South Pennines*, unpublished report for the Department of the Environment, London.

Hughes T G, 1996 *The Grey Slates of the South Pennines*, 2 vols, London, English Heritage, Derbyshire County Council and The Peak District National Park Planning Authority.

Hull, Edward. 1860, *The Geology of the Country around Wigan*. London, HMSO.

Jefferson D P, 1998 *Horsham Stone, West Sussex, the Potential Source of Paving and Roofing Stone at Broadbridge Heath*, unpublished report for English Heritage London.

Judd J W, 1875 *The Geology of Rutland and Parts of Lincoln, Leicester, Huntingdon and Cambridge*, Memoir of the Geological Survey, London, HMSO.

Kellaway G A and Welch F B A, 1993 *Geology of the Bristol District*, London, HMSO.

Kerr W J B, 1925 *Higham Ferrers and its Ducal and Royal Castle and Park*, Northampton, R Harris & Son.

Killmorgan Instrument Corporation, 1992 *Munsell Soil Colour Charts*, New York, Killmorgan Instrument Corp.

Knopp D, 1908 *Lancashire*, London, Victoria County Histories **2**.

La Touche W M D, 1923 Field Meeting – Plowden, in *Transactions of the Caradoc and Severn Valley Field Club*, **7**: 2, 61–3.

Lawson J, 1985 Harnage slates and other roofing materials in Shrewsbury and neighbourhood in the late medieval and early modern period, in *Transactions of the Shropshire Archaeological Society* **64**, 116–18.

Leary E, 1986, 1992 *The Building Sandstones of England*, Garston, Building Research Establishment.

Lebour G A, 1878 *Outlines of the Geology of Northumberland*, Newcastle-upon-Tyne, M & M W Lambert.

Lloyd N, 1929 *Building Craftsmanship in Brick and Tile and in Stone Slates*, Cambridge, Cambridge University Press.

Mackenzie E, 1827 *A Descriptive and Historical Account of the Town and County of Newcastle upon Tyne: Including the Borough of Gateshead*, Newcastle upon Tyne, Mackenzie and Dent.

Martin D J, 1986 *The Use and Conservation of Slate and Tile-stone Roof Coverings on Historic Buildings*. Unpublished MA dissertation in Conservation Studies, University of York Institute of Advanced Architectural Studies, York.

Mayhall J, 1861 *The Annals of Yorkshire from the Earliest Possible Time*, Leeds, Joseph Jackson.

Mills D A C and Hull J H, 1976 *Geology of the Country around Barnard Castle*, Memoir of the Geological Survey, London, HMSO.

Moorhouse S, 1990 The quarrying of stone roofing in West Yorkshire during the middle ages, in *Stone Quarrying and Building in England AD43 to 1525*, Parsons D (ed.), Chichester, Phillmore, 126–45.

Morton J, 1712 *The Natural History of Northamptonshire*, London, Knaplock and Wilkin.

Murchison R I, 1839 *The Silurian System*, London, John Murray.

Murchison R I, 1854 *Siluria, A History of the Oldest Known Rocks containing Organic Remains with a Brief Sketch of the Distribution of Gold over the Earth*, London, John Murray.

Neve R (T N Philomath), 1703 (1969) *The City and County Purchaser and Builder's Dictionary*, facsimile reprint Newton Abbot, David & Charles.

North R J, 1946 *The Slates of Wales*, Cardiff, National Museum of Wales.

Peate I C, 1940 The Welsh House, in *Y Cymmrodor, Transactions of the Honourable Society of Cymmrodorion*, **47**, 151.

Pevsner N B L, 1973 Northamptonshire, in *The Buildings of England*, London, Penguin.

Phillips J, 1848 *The Malvern Hills compared with the Palaeozoic districts of Abberley, Woolhope, May Hill, Tortworth and Usk*, Memoir of the Geological Survey 2 part 1, London, HMSO.

Philomath T N, see Neve R

Plot R, 1677 *Natural History of Oxfordshire*, Oxford and London.

Plymley J, 1803 *General view of the Agriculture of the County of Shropshire*, London, Richard Phillips.

Price J, 1995 *Stone Roof Coverings of the Cotswolds and West Oxfordshire*. Unpublished MA dissertation in Conservation Studies, University of York Institute of Advanced Architectural Studies, York.

Richardson L, 1904 *A Handbook to the Geology of Cheltenham and Neighbourhood*, Cheltenham, Norman, Sawyer and Co.

Richardson L, 1925 Certain Jurassic (Aalenian-Vesulian) strata of the Duston area, in *Proceedings of the Cotteswold Naturalist's Field Club*, **22,** 137–52.

Rudge T, 1813 *General View of the Agriculture of Gloucester*, London, Sherwood, Neely and Jones.

Salzman L F, 1952 (1997) *Building in England Down to 1540*, Oxford, Clarendon Press.

Saville R J, 1986 *Langton's Stone Quarries*, Langton Matravers Local History and Preservation Society 2nd ed.

Sharp H B, 1980 Collyweston stone slating techniques, in *Folk Life*, **18,** 28–37.

Sharp H B, 1982 Stone at Cambridge Castle: An early use of Collyweston Stone-slate, in *Cambridge Antiquarian Society*, **72**, 64.

Sharp S, 1870 The oolites of Northamptonshire, in *Quarterly Journal of the Geological Society of London*, **26**, 354–93.

Sorby H C, 1879 Address on the structure and origin of limestones: Anniversary Address of the President, in *Quarterly Journal of the Geological Society of London*, **35,** 56–95.

Stanley N 1993 *The Stone Slates of Derbyshire: An Appraisal of Farey's List* (unpublished report)

Stevenson W, 1812 *A General View of the Agriculture of the County of Dorset*, London, McMillan.

Sylvester-Bradley P C, 1940 The Purbeck beds of Swindon, in *Proceedings of the Geological Association*, **51**, 349–72.

Taylor W G, 2002 *Bacup Quarries* http://committed.to/bacuphistory

Thompson B, 1906 Northamptonshire, quarries and mines, in *Victoria County Histories*. London, Victoria County Histories **2** 298-303.

Thompson D B, 1993, The Grinshill Sandstones and Flagstones, in *Shropshire Naturalist*, **2**, 16-25.

Thompson, D B, 1995, *A Guide to the History and Geology of Quarrying at Localities along the Natural History Trail in Corbet Wood*, Grinshill, North Shropshire, Clive and Grinshill Conservation Committee.

Thorold Rogers J E, 1882 *A History of Agriculture and Prices in England from the year of the Oxford Parliament (1259) to the Commencement of the Continental War (1793), Vol IV: 1401–1582*, Oxford, Clarendon Press.

Toghill P, 1990 *Geology in Shropshire*, Shrewsbury, Swan Hill Press.

Traske C, 1898 *Norton-sub-Hamdon in the County of Somerset, Notes on the Parish and the Manor and on Ham Hill*, Taunton, Barnicott & Pearce.

Tuke J, 1800 *A General View of the Agriculture of the North Riding of Yorkshire*, London, McMillan.

Velacott H, 1908 Dorset, in *Victoria County Histories*. London, Victoria County Histories **2**.

Walton J, 1938 Some decadent local industries: quarrying, in *Transactions of the Halifax Antiquarian Society*, 41–59.

Walton J, 1940 Old Yorkshire quarrying industries, in *Quarry Managers Journal*, **Sept** 144–7, **Nov** 188–90 and 195.

Walton J, 1941 The stone-slate roofs of England, in *Quarry Managers Journal*, **June** 64–6, **July** 93–6, **Sept** 39.

Walton J, 1975 The English stone slater's craft, in *Folk Life*, **13**, 39–53.

Webster T, 1826 Observations on the Purbeck and Portland Beds, in *Transactions of the Geological Society*, **2**:2, 37–44.

Welch F B A and Trotter F M, 1961 *Geology of the Country around Monmouth and Chepstow*, London, Geological Survey of Great Britain.

West I W, 2001 *Purbeck type section, Durlston Bay, Swanage, England Part 2 Middle Purbeck Strata*. http://www.soton.ac.uk/~imw/durlmid.htm.

Williams J H, Shaw M and Denham V, 1985 *Middle Saxon Palaces at Northampton*, Northampton, Northampton Development Corporation.

Williams J H, 1979 *St Peter's St Northampton Excavations 1973–1976*, Northampton, Northampton Development Corporation.

Woodward H B, 1893 *The Jurassic Rocks of Britain Vol III The Lias of England and Wales (Yorkshire excepted)*, Memoir of the Geological Survey, London, HMSO.

Woodward H B, 1894 *The Jurassic Rocks of Britain Vol IV The Lower Oolitic Rocks of England (Yorkshire excepted)*, Memoir of the Geological Survey, London, HMSO.

Woodward H B, 1895 *The Jurassic Rocks of Britain Vol V Middle and Upper Oolitic Rocks of England (Yorkshire excepted)*, Memoir of the Geological Survey, London, HMSO.

Woodward J, 1728 *Fossils of all Kinds Digested into a Method Suitable for their Mutual Relation and Affinity*, London.

Wright W B, Sherlock R L, Wray D A, Lloyd W and Tonks L H, 1927 *The Geology of the Rossendale Anticline*, London, DSIR.

BIOGRAPHY

Terry Hughes worked in the Welsh slate industry at Penrhyn quarry for 10 years. He also has experience of the slate industry in the USA, throughout Europe and in India. Since 1983 he has represented the British slate industry on the British Standards Institution's committees for roofing slates (as chairman), the code of practice for slating and tiling and the National Federation of Roofing Contractors' slating and tiling committee. He has also been the chairman of the European Standards (CEN) Committee for slate and stone roofing products for 14 years. Since 1993 he has been an independent consultant. As English Heritage's consultant on slate and stone roofing he has carried out studies into stone roofing throughout England and has advised on the conservation and construction of the roofs of many historic buildings. As part of his work to support the regeneration of stone slate delving he established and chairs the Stone Roofing Association and the Stone Roof Working Group.

Sourcing new stone slates and re-roofing the nave of Pitchford Church, Shropshire

The stages of work and decisions that culminated in the re-roofing of the nave of St Michael's and All Angels Church, Pitchford, Shropshire

CHRIS WOOD *

English Heritage, 23 Savile Row, London W1S 2ET, UK. Tel: +44 20 7973 3026, fax: +44 20 7973 3130; email: chris.wood@english-heritage.org.uk

TERRY HUGHES

Slate & Stone Consultants, Ceunant, Caernarfon, Gwynedd LL55 4SA, UK

ABSTRACT

English Heritage has been campaigning to try to rejuvenate the stone slate industry, particularly in regions which now lack a supply of new material (English Heritage 1997a, Hughes 1997a). The aim of the campaign is to reverse today's dependence on inappropriate substitutes such as artificial slates, stone imported from other parts of the UK and abroad or the use of salvaged or cannibalised local material. This paper describes the decisions and actions that were taken to procure a new supply of stone slates prior to the re-roofing of the nave of Pitchford Church, Shropshire, where the local Harnage stone had not been produced for over 50 years. The project was implemented against a background of strategic research on stone-slating (Hughes, this volume) and included documentary research on the roof and stone, finding a source of new material and estimating the costs of production, obtaining consents, deciding the method of procurement, surveying the failed roof and detailing and completing the new covering. The church has been successfully re-roofed and lessons have been learned that may help others wanting to resuscitate former local roofing industries.

KEY WORDS

Materials procurement, stone slates, tilestones, delphs, delves, delving, quarry, quarrying, fissile sandstone roofing slates, Harnage slates, Hoar Edge Grit, re-roofing, Pitchford Church, Shropshire

INTRODUCTION

St Michael's and All Angels Church is a Grade I listed building (Fig 1) situated five miles south of Shrewsbury (OS SJ 527043). It lies some 50 metres from Pitchford Hall (Fig 2), also listed Grade I, which is regarded as one

Figure 1. The north elevation of St Michael's & All Angels Church, Pitchford, near Shrewsbury, in early 1998 (© Chris Wood, English Heritage).

* author for correspondence

Figure 2. Pitchford Hall showing its proximity to the church (scaffolded behind) during the re-roofing in July 1999 (© Chris Wood, English Heritage).

of the most important timber-framed buildings in England. By the early 1990s the nave roof of the church was in very poor condition and needed substantial repairs, so the Parochial Church Council (PCC) approached English Heritage for grant aid to help with the cost of the works. Some members of the PCC wanted the Harnage stone slates replaced with machine-made clay tiles to match those on the chancel roof. This was partly because these tiles would have been much cheaper, but they were also concerned that the roof might again fail prematurely; the nave having been completely re-roofed in 1937. The ability of stone slates to resist rain penetration was questioned.

At around the same time, repairs to the roofs of Pitchford Hall in recent years had had to be carried out in sandstone slates brought in from other areas of the country, [1] because of the lack of Harnage stone. By 1997 Shrewsbury & Atcham Borough Council were pressing the new owners to complete their roofing repairs. If a new supply of stone was to be won for the church, this would offer a timely opportunity to complete repairs to the Hall, using the correct material.

English Heritage were extremely keen to see the nave recovered in Harnage stone. Its *Roofs of England* campaign (English Heritage et al 1997, and Fig 3) specifically aimed to revive traditional stone slating which is such an important part of the distinctive appearance of so many areas and individual buildings in England (Fig 4). Over the twentieth century many roofs and much traditional craft knowledge have disappeared. As a result, repairs have often been carried out badly, either in alien materials brought in from elsewhere or, worse still, cannibalised or stolen from existing buildings. The character of many areas is increasingly being eroded as more roofs are lost.

Harnage slates provide a classic example of the problem. The stone has not been commercially quarried for over a century, so there are now a little over 20 buildings still covered in this unique material, although some of those remaining are very important historic buildings, such as Wenlock Priory (Fig 5) and Pitchford Hall. Local knowledge about the stone and the methods of detailing a roof has dwindled as few tradesmen have had the opportunity to work with these materials.

Figure 3. Front cover from the brochure of English Heritage's Roofs of England *campaign which sought to celebrate the nation's stone slate roofing traditions and draw attention to the current plight of the industry (© English Heritage Photo Library).*

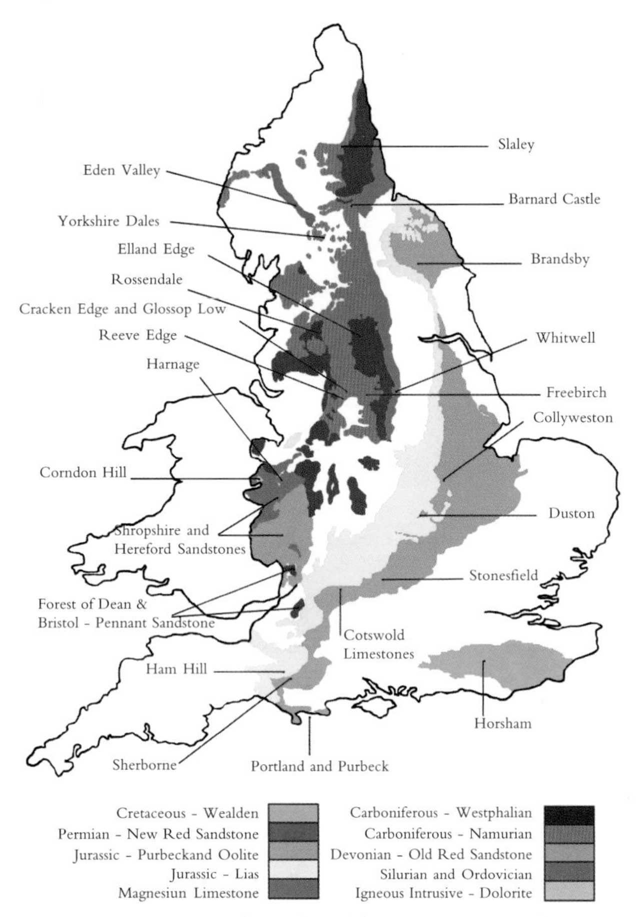

Figure 4. Map showing the geological formations important for stone slates in England and some historical locations for their production and use (© Terry Hughes, after British Geological Survey).

Re-roofing Pitchford Church in Harnage stone provided a unique opportunity to implement many of the objectives of the English Heritage campaign; in particular to assess the practical difficulties of finding and opening a new quarry to provide a source of stone. Studying the existing roof would establish why it had failed and would help with the detailing of the replacement. Although the immediate objective was to provide a supply of Harnage stone for Pitchford Church and Hall, the lessons learned could be very beneficial to those contemplating winning new sources of scarce roofing stone in other areas of England.

A number of sensitive issues had to be dealt with, not least the sometimes conflicting aspirations of those involved. Some members of the PCC were nervous about reusing stone slates and a balance had to be found between the desire for a faithful 'restoration' of historic detailing and the need for an effective and relatively maintenance-free roof. The listed church sits within a Grade II registered park as does the selected quarry site. In addition, the church bellcote was home to a colony of protected bats.

TERMINOLOGY

In this text, the term 'stone slates' is used throughout, although there are many other historical and regional names still employed in England for such materials. Some geologists prefer the term 'tilestones' to differentiate fissile limestone and sandstone roofing materials from true metamorphic slates.

HISTORY OF THE CHURCH

The name Pitchford is Anglo-Saxon in origin and derives from the crossing (ford) of the Row Brook which was contaminated with natural mineral oil (pitch) emerging from a local well (Plymley 1803, 70–1). The church is believed to have been founded and built by Ralph de Pitchford II around 1220 (Cranage 1903, 501–5). Architectural evidence points to an Early English date, although, interestingly, in the middle of the north wall are a few courses of crude herringbone-work which possibly are the remnants of an earlier structure. However, there is no documentary evidence of this.

Figure 5. Part of the roof of Wenlock Priory, Much Wenlock, Shropshire, another Grade I listed building still covered in Harnage stone slates (© Chris Wood, English Heritage).

There is some evidence that the timber roof dates from the sixteenth century (Stokes 1936), although this was mostly replaced earlier last century. Various alterations were made in the eighteenth century, including strengthening walls: 'On the north side of the church ... the wall has been thrust outward by the great weight of the stone shingles with which the roof is covered, and that it has had to be shored up by clumsy buttresses (Watkins Pitchford 1947). However, the church was not 'restored' until 1910. This included major works to the roof (see below) with further extensive repairs being undertaken in the 1930s.

The stonework is coursed red sandstone rubble, probably from the local Keele beds from one of the quarries along the Cound Brook (OS 550052). The grey ashlar dressings are thought to have come from the Grinshill Quarry some eight miles north of Shrewsbury (OS 524242). It seems likely that the church would have been externally rendered, and although minute traces of lime plaster have been found, evidence is inconclusive. Although there are references to stone shingles, 'slabs of shaley limestone' (Lawson 1985), on the roof going back to late medieval times, it is not known when these were first used, or indeed whether they were the original material.

Internally there are a number of features reflecting the historical development of the church, which still retains its ceilings. Most important are the monuments to Sir John de Pitchford (died 1285) and five incised alabaster slabs commemorating the Pitchford family of the sixteenth century.

REPAIRS CARRIED OUT IN THE TWENTIETH CENTURY

Annex 1 contains a detailed account of repairs carried out during the last century. New oak tie beams and king posts were inserted in 1910, when it is presumed that the chancel was covered in the Broseley clay tiles. It is probable that there was an insufficient quantity of reusable Harnage slates to repair the whole church. The condition of the roof was described as both dangerous and dilapidated in the 1920s and 1930s. Rafters and laths were severely decayed.

Figure 6. The state of the nave roof in 1995 where laths and pegs had rotted causing sections of the roof to fold inwards (© Chris Wood, English Heritage).

Messrs Treasure and Son began repairs in 1936 which amounted to a total reconstruction of the roofs (apart from the tie beams and king posts). New stone was extracted from Acton Burnell Hill (OS SJ 538007) and winning this turned out to be far more difficult than anticipated. Presumably the most accessible and suitably sized stone had already been extracted and much more overburden had to be removed than had been expected. The intention was to reconstruct the new roof as a replica of the old, for which the services of an architect were deemed unnecessary by the vicar. The work was duly completed but at a much greater cost than originally estimated.

An inspection in 1995 revealed more cause for concern (Fig 6). Water ingress had led to rotting of laths and pegs causing sections of slate to fold inwards. The cost of remedial works, which included renewing the ceiling and repairs to rainwater goods and drains was estimated and in January 1998 a grant of £76,200 was offered to the Parochial Church Council (80% of assumed costs).

Clearly, if Harnage slates were to be seen again on the roof, a suitable source of new stone would need to be found. English Heritage's Building Conservation & Research Team (BCRT) commissioned its slating consultant Terry Hughes, who had provided much of the research and technical expertise for the *Roofs of England* campaign, and Dr David Jefferson (see Appendix 2), its consultant geologist, to begin the search.

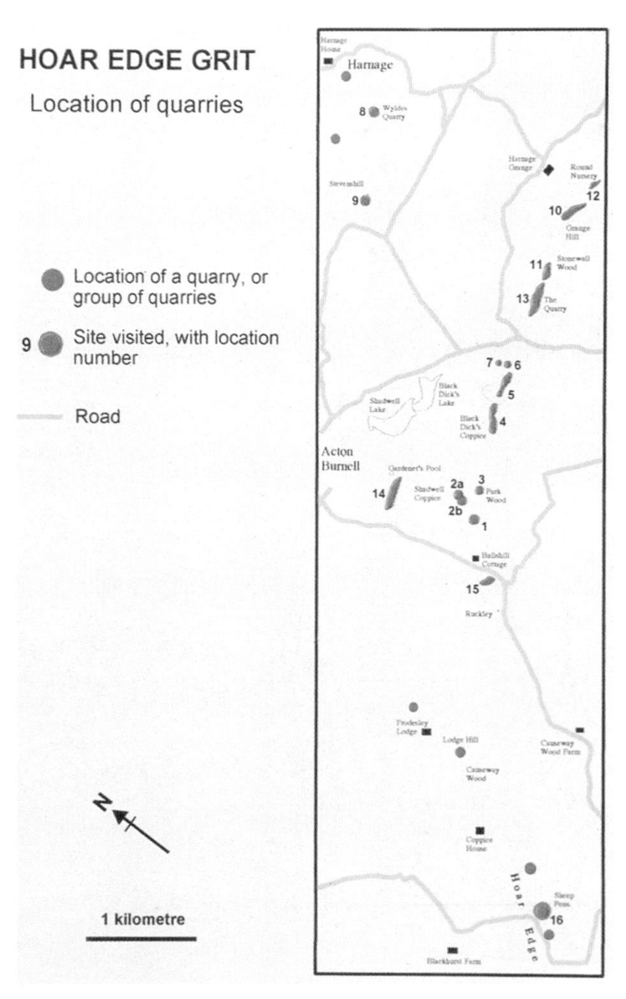

Figure 7. Historic quarries identified in the Hoar Edge Grit (© David Jefferson).

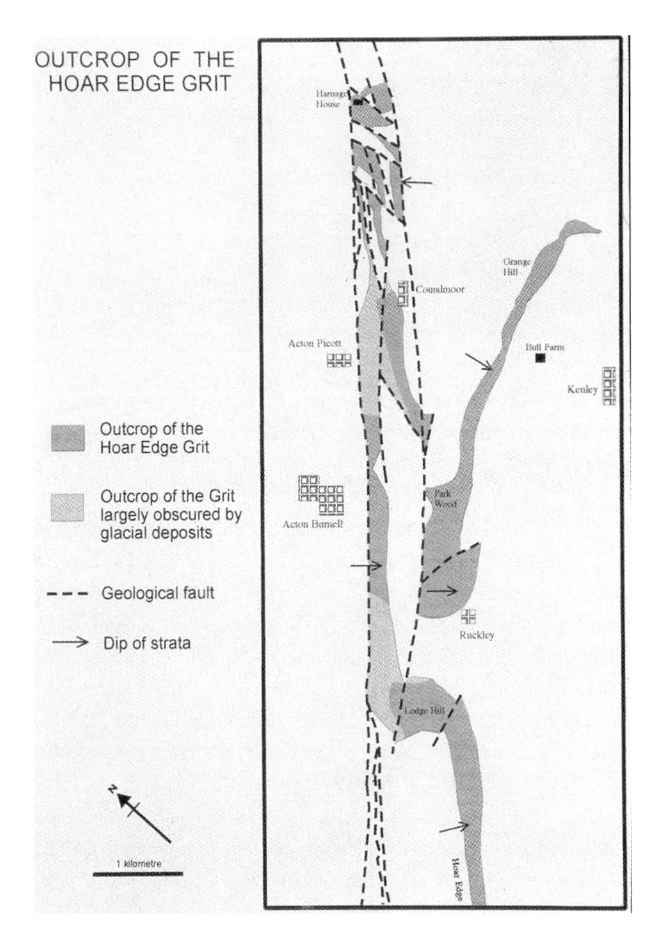

Figure 8. Diagram illustrating the lateral change within the Hoar Edge Grit, south-west of Acton Burnell (© David Jefferson).

HARNAGE STONE

Research carried out in the 1980s (Lawson 1985, 6) indicated that historical quarries lay along an outcrop of Ordovician rock lying south of Harnage Grange, on Grange Hill and in Shadwell Coppice (Fig 7). The stone is described as a hard, flaggy, calcareous sandstone with numerous shelly bands. The first mention of Harnage slates occurs in 1367 and 1368 when Robert de Harley, Lord of the Manor of Harley, sold the 'pieres de Harnage' from the roof of his kitchen and gatehouse at Harley' (Lawson 1985, 6), suggesting that the slates were already in use for high-grade buildings near the quarries by the mid fourteenth century. By the mid fifteenth century stone slates, very probably Harnage slates, were in common use in Shrewsbury.

A survey of the quarries in the Hoar Edge Grit which produced stone slates and other building materials was carried out by Hughes and Jefferson (Hughes 1997). A literature search, together with local information, enabled a large number of old quarries to be identified. Their study has confirmed that there is a lateral change (Fig 8) within the Hoar Edge Grit in the Acton Burnell area. Although the massive sandstones on Hoar Edge have been worked for building stone, a change in the strata to thin fine-grained sandstones and shelly sandstones in a north-easterly direction, has resulted in slate being extracted in the Harnage area.

Although the strata in the north-east of the outcrop have the potential to split into thin layers suitable for roofing material, it is only the naturally split weathered material close to the surface which has been used. Most of the outcrop suitable for use as stone slate has been worked in the past. However, three sites which might provide suitable stone were identified, although further investigations by excavating pits and trenches would be required to confirm this.

Geology

The stone from which the slates were obtained occurs in the Costonian Stage of the Caradoc Series. The strata of this age in this area are known as the Hoar Edge Grit, being named after the topographical feature where the rocks are most prominent.

In the region of Hoar Edge, the rocks consist of sandstone, grits and conglomerates, but there is considerable lateral change in the strata. North-east of Hoar Edge, the rocks become finer-grained and layers of shells, and in places thin limestones, occur in the sequence. There is also a thinning of the succession in this direction. This is because the exposed beds appear to have been formed along the original coastline. This environment of deposition resulted in sand-bars and spreads of gravel, as well

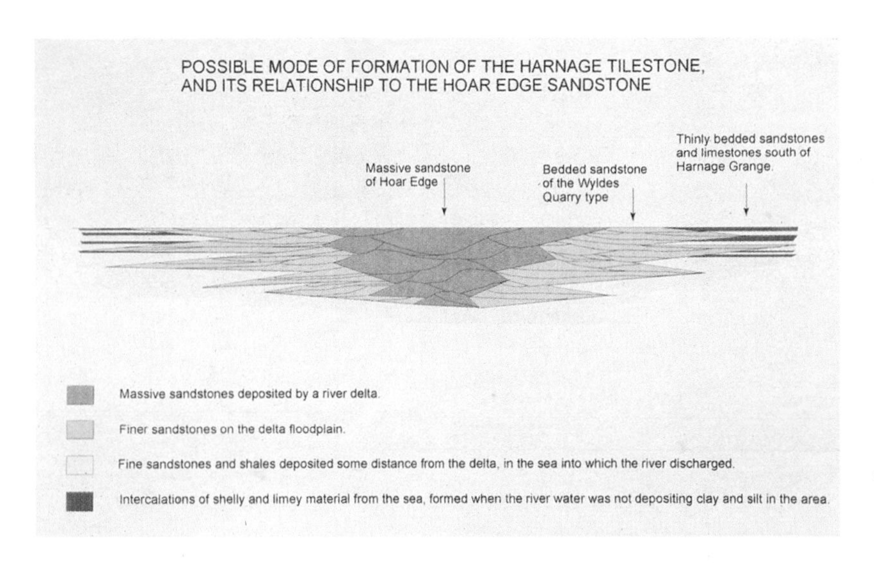

Figure 9. Conjectural explanation for the formation of Harnage slates and its relationship to the Hoar Edge sandstone (© David Jefferson).

as the presence of finer sediments in sheltered areas. It is also probable that rivers were depositing sands in the area in the form of small deltas. Figure 9 is a diagrammatic representation of how the massive sandstone of Hoar Edge may be related to the thinner, calcareous strata which are found on the ridge south of Harnage Grange, and from where the slates appear to have been obtained. At times of high river flow, or when coastal currents were strong, sands and gravels would be deposited; coarse material being laid down in the higher energy areas, fine sand sedimenting out where the water currents were slower. When the river or coastal currents had less effect, the marine environment would encroach into the deltaic or sand-bank environment, laying down fine sands or shell deposits, or even being suitable for the formation of limestone.

Compaction of the sediments by deep burial, followed by folding, faulting, uplift and erosion, has resulted in the distribution of the rocks formed, as they occur at the present time. The outcrop of the Hoar Edge Grit is shown on Figure 8, which also gives some indication of the complex faulting of the strata which has resulted in the rocks forming a series of discrete blocks, stretching from Hoar Edge in the south to Harnage in the north. The variation in the complexity of the faulting shown on the map is probably due to lack of exposure in some areas, particularly those where the rocks are covered by Pleistocene glacial deposits. It is very likely that the stone is broken up into small blocks by undetected faults in many of the areas where there is minimal exposure.

All the different types of stone in the Hoar Edge Grit sequence, whether massive sandstone or thinly bedded limestone, are hard and compact. Even the thinly bedded strata are compact when unweathered. Surface weathering in the period since the last glaciation has tended to weaken the rocks, especially the thinner bedded ones, along the bedding planes. This natural splitting of the strata is the reason for some of the stone being suitable for use as slate. However, this natural weathering feature only appears to extend down about two metres. Beneath this depth the stone is hard and the bedding planes tightly sealed. Whether or not the unweathered stone could be split by natural frosting or mechanical means was not known; it was, however, considered unlikely.

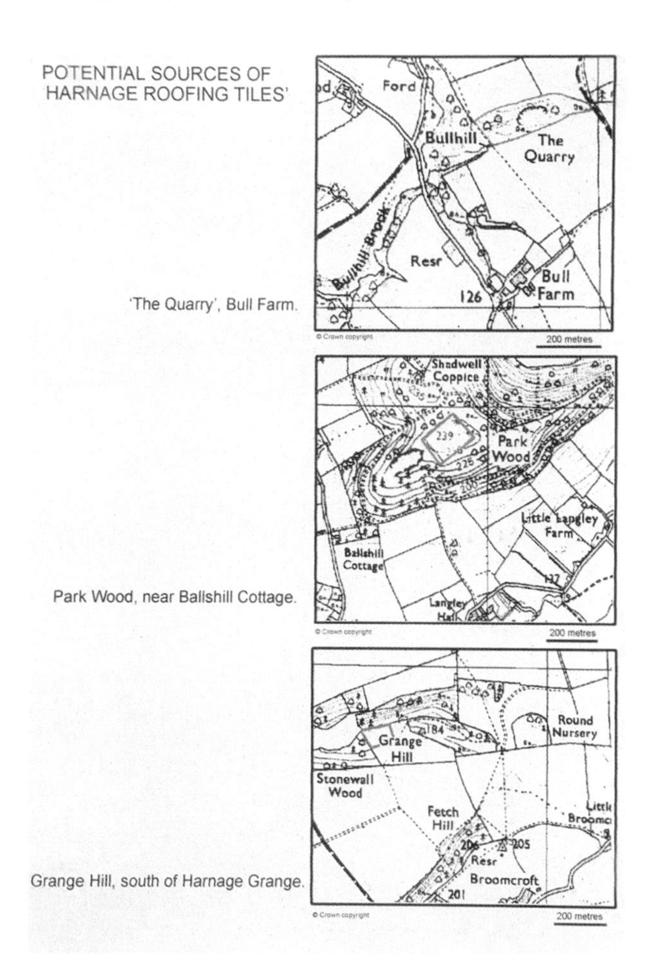

Figure 10. The three delves identified as being most likely to produce fissile stone (David Jefferson, based upon Ordnance Survey material by permission of Ordnance Survey mapping on behalf of the Controller of Her Majesty's Stationery Office © Crown Copyright GD03085G/03/01).

The quarries

A total of eighteen quarries, or groups of quarries, were identified from the available literature (Fig 7). [2] A further four areas were identified in the field, and local information provided one more. Seventeen groups of quarries were visited and enough information was obtained to obviate the need to visit the remaining six. The thinly fissile rock exposures were summarised as follows (see also Glossary below):

- At the west end of Park Wood, above Ballshill Cottage (OS SJ 5382 0154) (reported to be the source of the stone slate used for the re-roofing of Pitchford Church in the 1930s), a five-metre high face shows bedded, variably shelly, sandstone, typically 0.5 metres thick, dipping at between 25° and 30° in a southerly direction. The surface weathering zone is no more than about two metres thick. In this area the stone has split along bedding surfaces into flaggy pieces. It is estimated that about 20% of the volume would be acceptable for roofing purposes.
- The distribution of the different facies discussed above in the section on geology was confirmed. Thin, fissile sandstones, calcareous and shelly sandstones, together with limestones, are restricted to the area in, and to the east of Park Wood as far as Grange Hill.
- Potential sources of stone slates are, therefore, restricted to the area occupied by the 3.5 kilometre ridge within, and to the east of Shadwell Coppice and Park Wood. Furthermore, the bulk of the material suitable for use as a roofing material is restricted to the weathered zone of the bedrock, that is, about the top two metres. The solid material immediately beneath this may produce slates of a large size. However, if the bedrock was to be worked, appropriate equipment and personnel with the relevant skills in extracting building stone would be required, if wastage was to be maintained at an economic level.

Figure 11. A close-up of the nearby roof of Langley Chapel which shows the shelly and fine-grained sandstone slates (© English Heritage Photo Library).

Potential sources of new Harnage slates

Three areas were identified in February 1997 (Hughes 1997b) as being potential sources of stone slate. These are shown on Figure 10 and were reported as follows:

1. 'The Quarry', Bull Farm (OS SJ 5580 0154). This quarry, reopened about three years ago by the farmer, Mr John Wilde, may well provide large quantities of smaller tiles from the thick hillwash, as well as providing *in situ* stone, which may provide any large slates which are required. The quarry is situated in the central part of the region which was traditionally worked for slate.
2. Park Wood (OS SJ 5382 0088). An area of open ground, immediately to the east of the quarry which is reported to have supplied the roofing materials for Pitchford Church in the 1930s, appears to be unworked. The topsoil contains a large number of fragments of fissile stone. Although at the western end of the stone slate resource area, it is known that acceptable material can be obtained from this locality. The history of the site should be investigated, in order to ensure that it is not back-filled ground. If no history of quarrying is found, further investigations should be undertaken by excavating trenches across the site in a north-west to south-easterly direction.
3. Grange Hill (OS SJ 5632 0175). An area of open ground on the south side of the ridge exists immediately east of Stonewall Wood. Although the eastern end of this area appears to be underlain by shale, the western end, adjacent to Stonewall Wood, contains large quantities of fissile shelly sandstone in the soil. Trenches across this area, orientated in a north-south direction, would be required to determine whether or not any stone slate is present at this site.

Stone slate types

Two types of roofing stone are known in the Harnage area: a fine-grained sandstone and a shelly sandstone. There is also the possibility that an intermediate type may exist in which an individual slate may contain shell over only part of its surface. They have been mixed on some roofs. Langley Chapel (Fig 11) provides one example (although it is not known if this roof also includes other material imported from another region). Since both types have been found in the old quarries on Grange Hill it can be concluded that the mixture is authentic and does not represent use of imported material. The geology suggests that both types are available generally over the area but topographical and overburden problems may have restricted production.

There is no reason to assume that a mixture of rock types was exploited deliberately to produce a particular appearance. Presumably the mix on any one building would have been fortuitous and dependent on the availability of each type.

Exactly which slate was used on the roof of St Michael's and All Angels was not known at this time. If a mixture did exist then a decision would have to be made about trying to reproduce this for good conservation reasons, unless one of them was known to fail prematurely.

There was little to indicate that the apparently poor performance of the nave roof during the twentieth century was attributable to the slates themselves. Indeed, it was clear from historical evidence that Harnage roofing is a very durable material. The condition of the small number of slates which had fallen off the roof of St Michael's confirmed this. Observations of roofs in this part of Shropshire did not suggest that either type was likely to be significantly more durable than the other. Defined tests were not available to estimate the durability of individual stone slate types. Even adapting modern-day tests would not determine which of the two is the most durable.

Stone slates make a unique contribution to the distinctive appearance of regions. A clear understanding of the characteristics of the Harnage stone slates was needed to ensure that the methods of extraction and manufacture adopted would produce slates which would maintain the distinctive roofs still found in this part of Shropshire. Chris Harris of the Completely Stoned Company (CSC, now the Cotswold Stone Tile Company) (see Annex 2), who was to carry out this work was therefore closely involved with the roof investigation and the specification of the finished slate. [3]

The range and mix of sizes on the church roof was not known at this time. On other roofs, examples up to at least 600 mm^2 could be seen. Any reclamation of material from the existing roof would inevitably reduce the size range because of the need to dress off softened stone at the top edge. If the existing appearance was to be conserved, new slates of as large a size as possible would be needed.

The thickness of any roofing slate depends on the properties of the rock; in this case on the thickness of the bedding and the extent of weathering. There is a limit to the extent to which it can be controlled and ultimately a roof of thinner slates can only be achieved by rejecting the thicker examples with a consequential increase in cost. In practice, careful architectural detailing and roofing workmanship should ensure that a mixture of thicknesses will resist water ingress, so it was decided that the natural range should be accepted.

Since new slates would be obtained from the original beds there was no need to establish a control on colour. Similarly the surface texture would be an inevitable consequence of the rock used. The two types of Harnage stone would have dissimilar textures but the resulting mixture would nonetheless be authentic and desirable, subject, of course, to the findings of a detailed investigation into the existing mix of slates.

The full range of rock available from any source is likely to produce some slates which are excessively twisted or uneven. While competent stone slaters can be expected to deal with a wide range of slates by careful selection and placing, it must be accepted that there is a limit to what can be achieved without jeopardising the weathertightness of the roof. It was felt to be a wise precaution to seek the opinion of an experienced local slater during trial production to determine the suitability of a potential delph, but unfortunately no-one could be found, so this task was carried out by Chris Harris and Terry Hughes. The appearance of the edges of stone slates is vital to the appearance of the roof as a whole. Traditionally it arises from the method used to reduce them to size, which typically involves a tool similar to a brick hammer to trim the edge, often while it is held against a steel bar. Other methods such as the use of a cutting wheel or diamond saw produce a radically different appearance and should never be used as the final stage of processing (English Heritage 1997a). If such methods were to be used to reduce the rivings roughly to size, the edges would need to be subsequently hand dressed.

Conclusions of the investigation

There was good evidence that suitable sources of exploitable slates of both types existed in the area. Messrs Hughes and Jefferson recommended that trial excavations and production be carried out in at least one of the three most favourable areas (Hughes 1997b). Priority was to be given to the site at Park Wood since this was known to have been the source for the last re-roofing. However if both types of slates were required then it might have proved to be necessary to try one of the two potential sites on Grange Hill.

This uncertainty re-emphasised the need to investigate the roof of St Michael's and All Angels to determine whether both types of slate were present. If so, a decision was needed about manufacturing both types. It was agreed that during the inspection of the roof a preliminary specification for the preferred size range and thickness should also be prepared and an estimate made of how many slates would be required to make up those lost during stripping. This information would be especially useful during the exploratory investigations at the potential delves.

Regrettably, the roof could not be investigated until December 1998, a month after extraction ceased. This was because the contract to carry out the roofing work could not be placed until the costs and hence the grant were known and, consequently, the scaffold needed to investigate the roof could not be erected. This meant that there was little option but to accept the stone which was available. However, delving was closely monitored by Terry Hughes to ensure that suitable rock was being extracted and sent for processing.

Figure 12. Investigating a shelly stone bed at Bull Farm using a back-hoe from a local plant-hire firm (© Terry Hughes).

INVESTIGATION OF THE POTENTIAL DELVES

In September 1997 Terry Hughes carried out exploratory excavations in two potential sites with the agreement of the owners: Park Wood in the Acton Burnell Estate and in The Quarry at Bull Farm (Hughes 1997c). The objectives of these excavations were to:

- establish the suitability of the rock for the production of stone slates similar in size, thickness and appearance to those on St Michael and All Angels church
- estimate whether there is sufficient suitable rock to complete the re-roofing
- define the workable extent of the deposit of suitable rock as far as possible.

Site investigations

All trial excavations were carried out with a Caterpillar 428C back-hoe (Fig 12). A 600 mm bucket was found to be satisfactory for most of the work but in the unconsolidated hillwash at Bull Farm where the trench walls tended to collapse, a 1000 mm bucket was used. Excavations were carried out to a metre or so in depth, so shuttering was not needed.

The proportion of the excavated material which was potentially suitable for roofing was estimated visually. Samples were then split and dressed using hand tools to determine how easily it could be converted to roofing and to estimate the overall recovery from rock to product. At Bull Farm, where bedrock is exposed, rock was also extracted by hand using bars and chisels to confirm that larger slabs could be obtained and to compare the results with those won by mechanical excavation.

At Park Wood a series of twelve trenches were formed in a north-easterly direction from the old quarry to the crest of the hill (Fig 13). The excavations revealed that the fissile shelly material could be worked as a single bench with initial overburden and waste stock-piled to the north-west until space became available for back-filling.

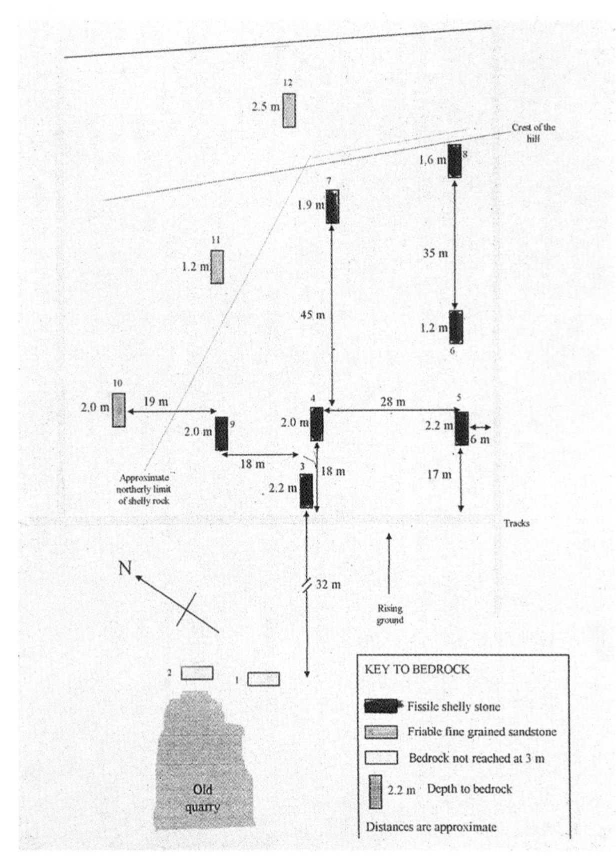

Figure 13. Plan showing the number of trial excavations that were needed to establish the material at the Park Wood site (© Terry Hughes).

It was estimated that the final back-filled ground level would be very similar to the existing, assuming that not more than 20% was removed for roofing and allowing for bulking-up of the waste.

Non-destructive survey techniques, such as radar, would not help in this sort of investigation. While it could detect a change in soil type (eg a gravel) or a horizontal break in a uniform granite, it would not be able to distinguish between a thinly bedded sandstone slate and a variably broken sandstone.

The investigations established that mechanical excavation would only be essential for the initial removal of overburden and establishing a working bench. However, a front-end loader or small back-hoe could be used to work the loose material without unacceptable damage to the rock. The bedrock below would need to be worked by hand tools to maximize the size of slabs and the overall recovery of useable rock.

The area underlain by fissile shelly rock was about 0.5 hectares in extent. Assuming a depth of 0.5 metres of which 10% would be suitable for roofing, it was estimated that there would be a volume of about 250 m^3 of useable rock equivalent to about 8,000 m^2 of roofing 30 mm thick on average. Assuming the coverage on the roof to be 35% because of double lapping, this would be sufficient to cover an area of about 3,000 m^2; more than enough for the whole roof of Pitchford Church.

At Bull Farm the ground in the quarry was much steeper than that at Park Wood and the loose material had slumped downhill to form a thicker cover over the bedrock. This had been removed over a substantial area during previous extractions, exposing bedrock of both shelly and fine-grained sandstone (Figs 14 and 15). Assuming 10% recovery from an accessible area of 2,500 m^2 by 1.5 metres depth, about 4,000 m^2 of roof could be covered, more than enough for Pitchford Church and to complete the re-roofing of Pitchford Hall. However to obtain large-size slates, the consolidated rock at a low level would have to be exploited. This would probably be more difficult and result in a lower recovery than from Park Wood, where the weathered, looser material was close to the surface. There is no set ratio in quarrying to decide the cost-effectiveness of removing a quantity of overburden to access a particular amount of slate. It is determined more by the potential output and its likely price.

Conclusions from the investigation

Sufficient suitable rock existed to re-roof Pitchford Church including the chancel. There would also be enough to complete the re-roofing of Pitchford Hall. The bulk could be obtained from the loose material close to the surface in Park Wood but some bedrock might need to be worked to ensure that the size mix on the existing roof was maintained.

At the quarry at Bull Farm the overburden was thicker and while it also contained disturbed shelly rock it was generally unsuitable for roofing, being too small and too

Figure 14. Bull Farm: roofing from Pitchford Church plus stone from the shelly bedrock at bottom left (© Terry Hughes).

Figure 15. Bull Farm: roofing produced from the shelly bedrock (lump hammer is 10 in long [254 mm]) (© Terry Hughes).

thick. It was underlain by fissile, shelly and fine-grained fissile sandstone bedrock and both types were exposed at the quarry floor. Elsewhere they were covered by up to five metres of overburden.

Both sources could be readily exploited using mechanical excavation to remove the overburden and hand tools to extract from the bedrock.

It seemed probable that most of the roofing material obtained in the past from the quarries along the two hills came from the beds of loose rock. This suggested that if they were reworked into the consolidated rock below there would be ample reserves to re-roof all the other buildings covered with Harnage stone as this became necessary. An opportunity therefore existed to extend the stone extraction latterly into previously worked ground to establish a stockpile of roofing stone for future needs and thus secure a supply of this unique type of roofing.

The roofing material from both sites was identical in appearance to that on the existing roof and if both loose rock and bedrock were worked the existing size mix could be guaranteed. However, because this rock produced very thick and uneven roofing slates, a balance would need to be struck between the cost of dressing the surface down to a smooth product acceptable to the roofer and the need to provide a technically satisfactory roof. This would be a fundamental cost factor and it re-emphasised the need to involve an experienced roofing contractor at an early stage of production. Ideally, to pre-empt disputes during the roofing about the quality of stone, the appointed slater would normally have been involved in defining and approving the slate specification. Unfortunately, as described above this was not possible and Chris Harris and Terry Hughes carried out this role.

It was clear from this investigation that potentially good supplies of stone were available, but many issues still needed to be resolved: consent from the land-owner to extract stone, planning permission, who would carry out the delving, and store and possibly stockpile material, how were contracts to be arranged and payments organised? Establishing costs would be difficult until a track record was established, as would quality control. The amount of new slates needed for the nave roof would be established once the investigation had taken place, but decisions would be required about any surplus that might result.

Negotiations with land-owner

Negotiations with the land-owner, the Acton Burnell Estate, were conducted through their agents, Cooke and Balfour (now Balfour, Burd & Benson) who were keen to see a remunerative use of the land, provided it did not interfere with the pheasant-shooting season which was due to start in November. Initially they were concerned that extracting stone could disturb the birds, but in the event the pheasants quickly adapted to the new activity.

The agents wanted to make their own decision about who should carry out the work subject to English Heritage's approval of their suitability. Competitive tenders were obtained from two companies experienced in the production of stone slates. The Tetbury Stone Company Ltd[4] provided a fixed price for the whole operation which included a sum for the land-owner. More of the risk and hence the profits would rest with the company. CSC proposed a cost-plus basis, in which they would be paid the costs of extraction and production and be paid a profit on top. On the basis of even the worst case estimate, the cost-plus option promised a considerable saving and as it was agreed to work on an 'open book' basis, the quantity surveyors could control costs as work progressed.

The application was made by the land agents with assistance from Terry Hughes and David Jefferson. Although this was submitted to the County Council, discussions were also held with Shrewsbury and Atcham Borough Council as a number of the policies in their Local Plan affected the proposal.

Planning and faculty applications

Discussions with Shropshire County Council minerals planners began in December 1997 and they had to weigh up the potentially conflicting objectives of conserving the natural and built environments. The Local Plan recognised the importance of authentic building materials and the range of proposed products was felt to be acceptable. However, the delph was to be hidden, if at all possible. This of course pointed to Park Wood, which although situated within a listed parkland, [5] was well screened by trees near the top of a hill.

The planners required information on:

- a needs assessment and justification including reference to pre-existing quarries in the area
- a method statement including stocking areas and overburden storage
- hours of operation
- local and environmental impacts and limitation proposals
- access and transport of products
- any potential pollution of water courses or supplies.

David Jefferson gave informal advice to the applicants on these operational details because of the novel nature of the planning application, particularly its very small scale which was quite unlike a normal quarry proposal. Subsequently CSC were selected for the delving which offered the clear advantage that the stone would be processed off-site in Wiltshire. [6] This would also confine vehicle movements to a few lorry-loads of the raw material and reinforced the message that what was proposed was in effect a very small-scale, temporary type of quarrying which would cease once the requisite amount of material had been won.

The faculty application [7] to the Chancellor of the Hereford Diocese and its diocesan advisory committee for permission to carry out roofing works at St Michael's, was made by Arrol & Snell (A&S), the newly-appointed architects to the PCC in July 1998. This included an assurance by English Heritage that if it proved impossible to produce stone to recover the roof, then favourable consideration in terms of grant aid would be given to use clay tiles to match the chancel, subject to the approval of the local authority.

Facilitating the delving

Although the faculty had been agreed and planning permission granted there were many issues that needed resolution before work could start. The contractor was not going to begin until a contractual order had been received. English Heritage had offered a grant for the re-roofing based on an earlier estimate produced by their commissioned architect, who had no information on the actual costs of extracting the stone. English Heritage's normal procedure is simply to offer a percentage grant for the cost of repairs with the applicants being responsible for materials and workmanship. In this instance it was the PCC who would be the applicant and, through their architects, responsible for instructing CSC. Understandably, the PCC were unable to do this, because of uncertainties over costs and the amount of stone needed. At English Heritage decisions also had to be taken about the purchase of the spare stone slates; a novel exercise since its activities are normally confined to providing grants for others to purchase materials rather than supplying them themselves. Issues such as the value of the surplus slates, how this would be described in the English Heritage accounts and product liability all had to be considered. Fortunately, the objective of tackling the fundamental problem of conserving buildings for which materials are not available met with support and the proposals were accepted.

Using figures supplied by Terry Hughes and CSC, a number of assumptions were made in order to develop a cost estimate. The roof area (including bellcote) taken from drawings was 230 m^2 which equates to 24.75 roofing squares (1 square = 100 sq ft [9.3 m^2]): say 26 squares to allow a margin of safety. Based on the weight of fallen slates from the church it was assumed that they would weigh two tonnes per roofing square. Based on experience during the investigation works, CSC estimated that the recovery would be between 15 and 25% of the ground worked and that there would be a processing loss of 40% to 60%. Between 350 and 870 tonnes of ground would have to be worked.

This gave a range of costs to which the setting up and reinstatement element of £8,500 was added. The costs of transport were included, together with a profit for this part of the operation. A similar exercise for the conversion to finished slate gave the range of possible costs of production. Although ultimately 26 squares might not be required, producing any less would greatly increase the unit cost. Producing more would, of course, reduce it.

The total costs of the work would include the cost of reslating the roof, which would be more expensive if most of the existing slates were to be reused, because of the time taken in selecting, sorting and redressing. Therefore a series of costs were calculated from the most favourable i.e. least amount of selection wastage and maximum use of new slates, to the worst, which included maximum wastage and 50% reuse of existing. For 26 squares this gave a range of £55,700 to £79,350. These figures would of course be reduced if an order was received from the owners of Pitchford Hall as this would have introduced greater economies of scale.

After much deliberation and rehashing of figures, approval from English Heritage's Cathedrals and Churches Advisory Committee, now the Places of Worship Committee, was given for increasing the grant by £69,282, the total amount for the roof works being £79,350. English Heritage calculated a 90% grant based on the assumption that the whole of the nave roof would require new stone, but adding in the additional projected labour costs for re-using 50% of the existing slates. Any surplus would become the property of English Heritage, on the assumption that material reclaimed from the roof would be used first, in accordance with good conservation practice. Unused new material would be used as a small stockpile for the church and for English Heritage properties roofed with Harnage (such as Langley Chapel), with the rest being intended for use on repairs to other local historic buildings. Committee approval had to be sought for this unusual procedure and A&S were authorised to instruct CSC on 12 October 1998 to commence working on a cost plus basis.

There were a considerable number of risks involved with the project. Although the trials indicated that there was sufficient fissile stone, there was no certainty that the right quality and quantity would be produced. Nor was it known how much wastage would really occur. The advantage of the increased grant, however, did mean that, provided all the works were carefully monitored and regularly reviewed, a decision could be made to cease the delving and still retain sufficient of the grant to roof the nave in clay tiles. Now everyone had to hold their nerve.

THE DELVING AT PARK WOOD

English Heritage and the PCC appointed quantity surveyors (see Annex 2) to closely monitor progress at the delph because of the uncertainties about production and costs. The total budgeted maximum for extraction and production was £47,268. Work was to progress uphill from the south-west edge of the site to permit drainage away from the working area but this resulted in an unpromising start. Before suitable fissile rock was exposed delving had progressed along the southern edge to the southernmost corner of the consented area. This did not conflict with the trial excavations as it was known that more productive areas would be found uphill. Fortunately, after further work, the fissile rock was found at 1.4 m (4 ft) depth. Until this point, in an effort to avoid damaging the rock a small excavator (Fig 16) had been used, but because of time constraints a larger one similar to that used in the trial excavations was commissioned to get to the more productive rock.

Working conditions for the three quarrymen were extremely bad in October 1998. It rained for much of the month, indeed it was the wettest October in Shropshire since 1841. Apart from the personal discomfort the steep access track became impassable which meant that the slab which had been put into the crates could not be moved down the hill for despatch to Wiltshire for processing. Eventually a longer, less steep track was used and a tractor and trailer hired at additional expense to remove the crates.

Fortunately, both the Acton Burnell Estate and the Shrophshire planners were sympathetic to this predicament and extensions of time were permitted. The Estate found that, in fact, the working created little disturbance to the pheasants which fed happily among the excavators, so although operations ceased some days before a shoot, delays were comparatively insignificant. The planners also agreed the move into Phase II of the site, accepted the need for oversized bunds and allowed working into December.

Within two days a bench had been exposed at about a metre (3'3") depth and was producing 30 mm (1¼") thick slabs up to two metres square. In fact, some slabs were so large they had to be broken up before lifting. This was just what was wanted: as the roof investigation was to show, larger slabs would be most in demand. At the peak of production three excavators were employed to remove the overburden, loosen the fissile rock and carry rock and packed crates. Most of the working of the slate beds was with hand tools, bars and bolster chisels (Fig 17). A hand-held cutting wheel was used to roughly size the largest slabs for transporting. Although progress was now promising, there was always the danger that time would be lost searching for the next working bench. To avoid this, some excavation was carried out in adjacent ground so that an alternative location would be available immediately if the fissile rock ran out. One tonne of rivings were taken down to Wiltshire for trial processing. The outcome was very encouraging. Initial wastage of 50% quickly fell to 40%: the low end of the cost estimate. This promised a reduction in the number of lorry-loads of stone that would have to be carried to Wiltshire from six to four.

By the end of the second week CSC transported 17 tonnes to Wiltshire and, most gratifyingly, the low level of wastage was confirmed. Over six squares of finished slate were produced. On this basis only 74 tonnes would be needed to produce the 26 squares needed. The total production cost appeared to be about £590 per roofing square, mid-way between best and worst estimates.

Once the bench was exposed both it and the rivings were kept wet (easily done in torrential rain), because the rock is easier to cleave in this state, and covered with polythene at night to minimise the risk of frost damage. Although some stone slates have traditionally been split through the action of frost (such as Collyweston and Stonesfields), no-one knew the potential effects on Harnage stone. Rock when it is first quarried contains sap, and when it dries the minerals change and this can prevent the stone from being split along its bedding planes. For these reasons, the crates of rock were wrapped in cling film for the journey to Wiltshire (Fig 18). After cleaving, the rivings were allowed to stand for a day or two because dry rock is easier to dress.

Figure 16. A disappointing start to the work with a relatively small amount of usable stone excavated, prompted the hire of the larger excavator to remove the overburden. This photo shows the small excavator (© Chris Wood, English Heritage).

Figure 17. Stone extraction was still mainly done by hand tools, bolsters, hammers and bars, to roughly produce the thickness required for stacking and transporting (© English Heritage).

Figure 18. Roughly prepared slates, being protected from the weather by polythene prior to despatch to Wiltshire (© Chris Wood).

Figure 19. Dressing of the stones using hammers, under cover in Wiltshire (© Chris Wood).

Figure 20. The access scaffold for the roof investigation with three levels: the eaves, 2 and 4 metres above (© English Heritage Photo Library).

Figure 21. The scaffold needed to be sturdy to accommodate the equipment and the slates removed during the investigation (© English Heritage).

As production moved up the hill from the southern corner, productive rock was found some 450 mm (18") below the surface as predicted by the trials. Amended calculations of the amount of stone needed for a square of slates meant that sufficient stone had been won for the instruction to be given to stop work on 27 November, although CSC carried on in order to win 10 squares for Pitchford Hall. The daily cost of delving was £580.

In all approximately 90 tonnes of stone were delivered by articulated lorry to Wiltshire and processed by

Figure 22. The accumulation of moss and lichens on the northern slope of the nave roof obscured the edges of the slates (© Terry Hughes).

hand. Diamond-tipped saws were used to roughly shape the slates and these were hand-dressed using hammers (Fig 19).

There were other advantages in carrying out the final dressing of stones in Wiltshire, besides the local environmental ones. For CSC it was considerably cheaper for their men to spend four to five weeks in Shropshire extracting the stone and then return to Wiltshire to help others with the dressing, than trying to complete the whole job on-site. The worry about security was removed, the work could be done under cover and the problems of noise and traffic movements around Park Wood were kept to a minimum. Finally, it meant that CSC could also record exactly how many finished slates of various sizes it had, which in theory would make the planning of the new roof much easier.

In total 90 tonnes of stone produced 62 tonnes of finished slate. The average, daily production was approximately five tonnes. The next stage of the project was to determine exactly the quantity and sizes of new slates that would be needed. A close examination of the existing slates was needed and this would help to determine why the roof had failed in such a relatively short space of time.

INVESTIGATING THE NAVE ROOF

Method

The roof construction was investigated by Terry Hughes on 1 and 2 December 1998 (Hughes 1999). The investigation covered an area of about two metres width from eaves to ridge on both sides of the roof and was carried out from a scaffold at three levels: at the eaves and at approximately 6' (1.8 m) and 12' (3.6 m) above (Figs 20 and 21). This allowed all areas of the roof to be reached for careful dismantling by hand. The main area of the scaffold at the eaves was larger than normally required for roofing as it would later be used for the main repair work when adequate space would be needed to sort and redress the existing slates. Some of these were over 600 mm (24") square, the largest weighing up to 1 cwt (50.8 kg). The whole area under investigation was covered and weather-

Figure 23. Measuring the slate lengths, laps and margins on course 9 (© Chris Wood, English Heritage).

Figure 24. Inadequate head and side laps on the north slope result in water getting into the roof (© Chris Wood, English Heritage).

proofed with scaffold sheeting. To ensure adequate lighting for photography (plastic sheeting gives photographs a green cast) and late working, two portable tripod mounted floodlights were used.

The plant growth (lichens, mosses and ferns) was so thick over most of the roof that the vertical and bottom edges of the slates were often obscured (Fig 22). The stone slates were therefore cleaned off with a scraper and stiff hand brush prior to progressive stripping from the ridge. As each course was removed the exposed area below was photographed using 400 ASA monochrome and colour negative film. The photographs were taken from a point perpendicular to the slating wherever the scaffold permitted and included a scale rule with 50 mm (2") divisions (Fig 23) and a number 40 mm (1¾") tall for each course. Groups of slates were also photographed laid out on the scaffold to record their shape, degree of shouldering and the ratio of the length to the width. All these factors are important in preventing the penetration of rainwater. A description of features of each course were recorded into a dictaphone as work progressed.

The slate lengths below their fixing holes, head laps and margins, the vertical dimension of the exposed area of the slate, were measured at each course (Figs 23 and 33), except where the courses had collapsed onto each other. In this situation only the slate length was meaningful.

On a roof of this type and especially because of its deteriorated condition, measurements of head laps and margins cannot be precise. The margins varied considerably along each course. Also, the head lap was somewhat theoretical. It is a dimension by which the roof is set out rather than the precise head lap achieved which is affected by the shape of the slate's top edge and the position of the fixing hole. The head lap achieved depends firstly on the position of the perpendicular joint in the course above in relation to the curve of the slate head (if it drops below the peg hole) and, secondly, on the position of the peg hole laterally in relation to the perpendicular joint. For these reasons it is not possible to work out any of the three measured dimensions from the other two.

In spite of these difficulties, with care and a large number of measurements, it was possible to draw conclusions about the construction of the roof, and the apparent intentions of the contractors.

Roof condition

The church stands in a very sheltered environment and is surrounded by tall trees on all sides. This would have reduced the roof's ability to dry out after rain and encouraged the growth of lichens, mosses and ferns.

The roof area investigated was entirely formed of Harnage stone slates hung with oak pegs on oak laths and mortar bedded, except over the rafters where ferrous nails had been used for fixing. The bedding virtually filled the space between laths with mortar and in places this extended almost to the tail of the course above. The rafter pitch was 50°. The laths varied from about 64 x 10 mm (2½" x ½") at the eaves to about 30 x 10 mm (1¼"x ½") in the upper courses. The composition of the mortar was not investigated, although its constituents and texture suggested it might have been 1:2:9 cement:lime:aggregate. There was no provision for a flow of air through the rafter space so any dampness below the roof surface could only escape through the plastered ceiling or the slates.

The stone slates were generally small. Slates 300 mm (12") long or less accounted for 82% of the courses on the north slope and 85% of the south. The largest slates were 550 mm (22") long on the north slope and 650 mm (26") on the south. There were examples of very poorly shaped slates which would certainly have resulted in water penetration, but in general they were not excessively shouldered. There were many instances where slates which were wide enough to be used but because of the setting out of the roof, had far too small a side lap (Fig 24).

The holing of the stone slates was mostly mechanically drilled, about 30% being hand pecked. This suggests that in 1936 almost a third (the hand pecked) were reclaimed from the previous roof. Some of the slates had been reshaped for re-use, with some being turned through 90°. This did not appear to have weakened them. [8]

At the top of the slopes the slates lay at a pitch of only 10° to 18° which would have increased the amount of water penetration. Since the pitch of slates is dependent on their thickness, length and head lap it is inevitable that

Figure 25. The growth of lichens, mosses and ferns, particularly on the north slope, helped to retain moisture in the roof and restricted its drying out (© Chris Wood, English Heritage).

it will tend to reduce for the smaller slates towards the top of any roof slope. Elsewhere on the church, their pitch was generally about 30°.

Many of the laths had bowed down the slope, because they were too thin for the weight of the stone-slates and because of the wide spacing of the rafters: 685 mm (27"). This resulted in a reduction of the head laps. At the upper courses where the laths were small, 30 mm x 10 mm (1¼" x ½"), this was likely to have occurred early in the life of the roof. However most of the bowing had been caused by the laths rotting as a result of fungal attack brought on by moisture ingress.

The plants which had established on the roof, especially the ferns on the north slope, had dammed the flow of water down the perpendicular joints, increasing the extent of water penetration between the slates (Fig 25). Their presence would have also reduced the speed and extent to which the slates and the rafter space could dry out. The fern roots had also spread between the slates lifting them, allowing more water to penetrate, further aggravating the drainage problems (Fig 26). All these factors had created and maintained wet conditions at the laths and rafters.

The north slope was wet and substantially deteriorated. Over a large area the laths and pegs were rotted and eleven courses in the middle section had completely collapsed. Consequently the slates here had slid down the slope, reducing the head lap and were also resting on the ceiling. Here it was not possible to measure constructional details. In the wettest areas the rafters had been slightly affected by rot and the battens supporting the ceiling had rotted completely, leading to its collapse. Subsequent repairs had cured the symptom but not dealt with the cause. The lowest courses, where the slates and head laps were larger, were in much better condition and generally drier. Bedding the heads of the slates had not been necessary here so the wicking effect was avoided. On the north slope the extent of badly rotted laths coincided approximately with the courses of 250 mm (12") slates and 75 mm (3") head laps and accounted for 82% of the slope.

Figure 26. Root growth under the slates which together with the mortar bed helped to retain moisture, contributing to the rotting of the oak laths and pegs (© Chris Wood, English Heritage).

The south slope was in better condition and although the laths and pegs were damp and weakened they had not rotted away as on the north slope. Constructional details (laps, mortaring etc) were essentially the same as those on the north slope and would have suffered a similar degree of water penetration. This indicated that the south slope had dried out more quickly or thoroughly than the north.

There were fewer plants on the south slope, especially ferns. This was partly due to the drier conditions which would have restricted their growth. In turn their absence would have allowed faster drying below the slates.

Measured head laps varied between 100 mm (4") and zero. It appears that the general intention was to set out the roof for a 75 mm (3") head lap in the lower courses, reducing to 50 mm (2") towards the ridge (the correct principle is that the head lap should not exceed one third of the slate length). Adequate head laps had not been achieved over much of the roof. Possibly the difficulty in obtaining a good supply of stone slates in 1936 may have led the roofer to reduce the head laps in order to cover the whole roof. Sadly, this might have been avoided had the church been able to afford to employ someone who understood the correct method of slating to supervise the work.

Summary of reasons for failure of nave roof

The immediate cause of failure was the rotting of the laths and pegs brought on by the persistent wet conditions below the stone slates. The following factors had contributed to this failure:

- inadequate head laps due to poor setting out of the coursing, possibly due to insufficient slates to cover the roof area at the appropriate lap
- bowing of laths
- inadequate side laps, partly due to poor setting out and/or inadequate supplies of suitably wide slates
- the generally small size of the slates which, because of their curved top edge, further reduced the head and side laps. The smaller the slate the more extreme the reduction.
- excessive mortaring, as the mortar would have held any moisture, exacerbating the wetness and rot. In the worst cases the mortar was so close to the exposed

edges of the slates that it would have drawn moisture into the roof.

- plant growth on the roof. This had had the effect of reducing the flow of water down the roof and damming the drainage down the perpendicular joints and increasing the spread of water between the slates and into the lath space.
- plant roots between the slates, especially ferns. These had had a similar effect to the mortar in creating and maintaining a damp environment. (Plant transpiration must also to some extent have resulted in moisture removal during dry periods.) The fern roots had also forced slates apart increasing the likelihood of water penetration.
- condensation of moisture from the interior of the church within the roof structure. There was little or no ventilation out of the body of the church with a very high water table below.
- lack of ventilation through the rafter space and through the slating
- the sheltered environment around the church which would have restricted drying of the roof, especially the north slope, and, consequently, encouraged plant growth.

Suitability of slates for re-use

Generally the weak mortar adhering to the slates was easily removed without damage. Most of the slates were sound despite the wet conditions they had endured for the last 60 years (or more in the case of the previously reused slates). This confirms that Harnage stone slates are very durable and as many of them were lost in 1936, it suggests that the pre-twentieth-century roof may have been of a great age.

Although a good proportion of the stone slates were sound enough to be reused, the investigation clearly indicated that if the new roof was to resist water penetration, smaller slates (say less than 325 mm [13"]) should be avoided except on the highest courses. Any reused slates would need to be carefully inspected and dressed to ensure they were free of physical defects and had a suitable shape to provide good side laps.

Reconstruction

Inspection of similar roofs at Wilderhope Manor and Langley Chapel (OS 126, SJ 538001) indicated that there is nothing inherently unsuitable about Harnage stone slates which would require extreme or special precautions in reconstructing the church roof. Indeed, provided that the pitch is steep, 50° is satisfactory, the head and side laps are adequate and the construction does not promote condensation or moisture retention, the roof should have a very long life.

Therefore the primary technical objectives for the design and reconstruction were to:

- minimize the penetration of water through the stone slating
- minimize as far as possible the amount of moisture from within the church passing into the rafter space and condensing
- adequately ventilate any moisture which does get into the roof
- establish a maintenance programme to ensure these objectives are not compromised subsequently

The results of the investigation clearly focused minds on these important technical issues, even though the aim was to avoid any change to the appearance of the roof and respect traditional local detailing.

DESIGNING THE NEW ROOF

Clearly it was imperative to design a roof which would perform its primary function of keeping the water out, in a very damp environment. A&S prepared their specification which was subjected to extensive consultation and debate, particularly over the use of boarding, felts, fixings, ventilation and the detailing of head and side laps.

The first issue to decide was whether to adopt the original detailing, which would have seen random slates hung by oak pegs in diminishing courses on riven oak laths. From a philosophical viewpoint, this was the correct approach to adopt. This traditional method had worked well, as current examples nearby at Langley Chapel and Stokesay Castle showed. However, neither of these two buildings are occupied or have any form of active use as they are Scheduled Ancient Monuments in the guardianship of English Heritage. While there was general confidence that many of the faults of the 1930s would not be repeated, a number of factors militated against this purist approach.

Some members of the PCC were unhappy about historic detailing which had already failed prematurely. Improvements obviously could have been made, including the use of a secondary lath or batten above the peg to prevent it twisting, nailing rather than hanging the stoneslates, or the use of non-corroding fixings for the laths and the traditional use of torching to the underside of the slates. However the unevenness of the slates raised concern that water ingress is always possible, and torching, while often effective, needs maintenance. This could only be done by removing the ceilings which was aesthetically and technically unacceptable (Fig 27).

The church would still be subject to a very damp atmosphere even though improvements were made to the environmental conditions within the church. It is only used once a fortnight and occasionally heated, so the inherent damp associated with a high water table means that high internal humidity will be a constant factor. Condensation at roof level would therefore always be a risk.

A&S were keen to use horizontal boarding which had been used successfully by the noted Victorian architect George Devey in his restoration of Pitchford Hall. Diagonal boarding was the obvious method which offered a number of advantages; the most important being that it could provide wind-bracing which the roof lacked and also offer greater flexibility for the gauging of the

Figure 27. The ceilings would have had to be removed if the riven laths and torching were to be included as part of the detailing of the new roof (© English Heritage Photo Library).

Figure 28. The horizontal boarding cut to accommodate the asymmetrical plan shape of the nave caused by the unrestrained thrust of the roof in earlier years (© Chris Wood, English Heritage).

Figure 29. Full scaffolding and sheeting protecting the church during the works (© Chris Wood, English Heritage).

slates (it would avoid a gap between the boards coinciding with a line of pegs or nails). The arguments against diagonal boarding were that the nailing point could also coincide with a gap even on diagonal boarding (although a new fixing hole could always be made) and ends of the diagonal boards at the eaves and ridge might coincide with a rafter. This is normally overcome by using noggins to support the board ends. In the end horizontal boarding was used and the width of the boards were adjusted to suit the variable gauge of the slating. The boards were cut to fit the plan shape and the change of pitch at the eaves, which had been brought about by the displacement of the wall heads resulting from the unrestrained thrust of the roof in earlier years. This can be seen beyond the diagonal cut in the boards in Figure 28.

The final design also had to overcome the concerns of the PCC caused by past performance and the consequent lack of confidence in a stone slate of very uneven size and thickness. This meant minimizing any possibility of ingress by rain, as this would not only exacerbate the inherently damp conditions in the church, it would also be expensive and disruptive to rectify and could require the replacement of the plastered ceilings.

There was complete agreement to extend the scaffold to provide a fully protected working platform, capable of accommodating the weight of the stone (Fig 29).

Specification

A&S produced a very detailed and comprehensive specification and schedule of works against which invited main contractors could tender. The design was for the roof to be horizontally boarded in new tanalised 32 x 200 mm (1½ x 8") sawn boards in Douglas Fir, rough side up, with 5 mm (¼") air gaps. All slopes were to be covered with heavy-duty reinforced bitumen roofing felt (Ruberoid Zylex) fixed with 40 mm (1¾") copper clout nails. This was to be lapped over the ridge, and punctured for the ventcaps. Softwood laths were to be fixed to the side of the rafters to provide a 50 mm (2") airway and to carry the between rafters ceiling (Figs 30 and 31). All slates were to be nailed with 15 x 5 mm (¾ x ¼") or longer stainless steel angular ring-shanked nails, the length depending on the kick at the eaves or the slate thickness.

This detail was later amended to include a plastic sarking felt as a separating layer below the bituminous felt, which was included to prevent the bitumen melting and sticking the felt, which could lead to long-term horizontal splitting. This would also allow a heavier felt to be used

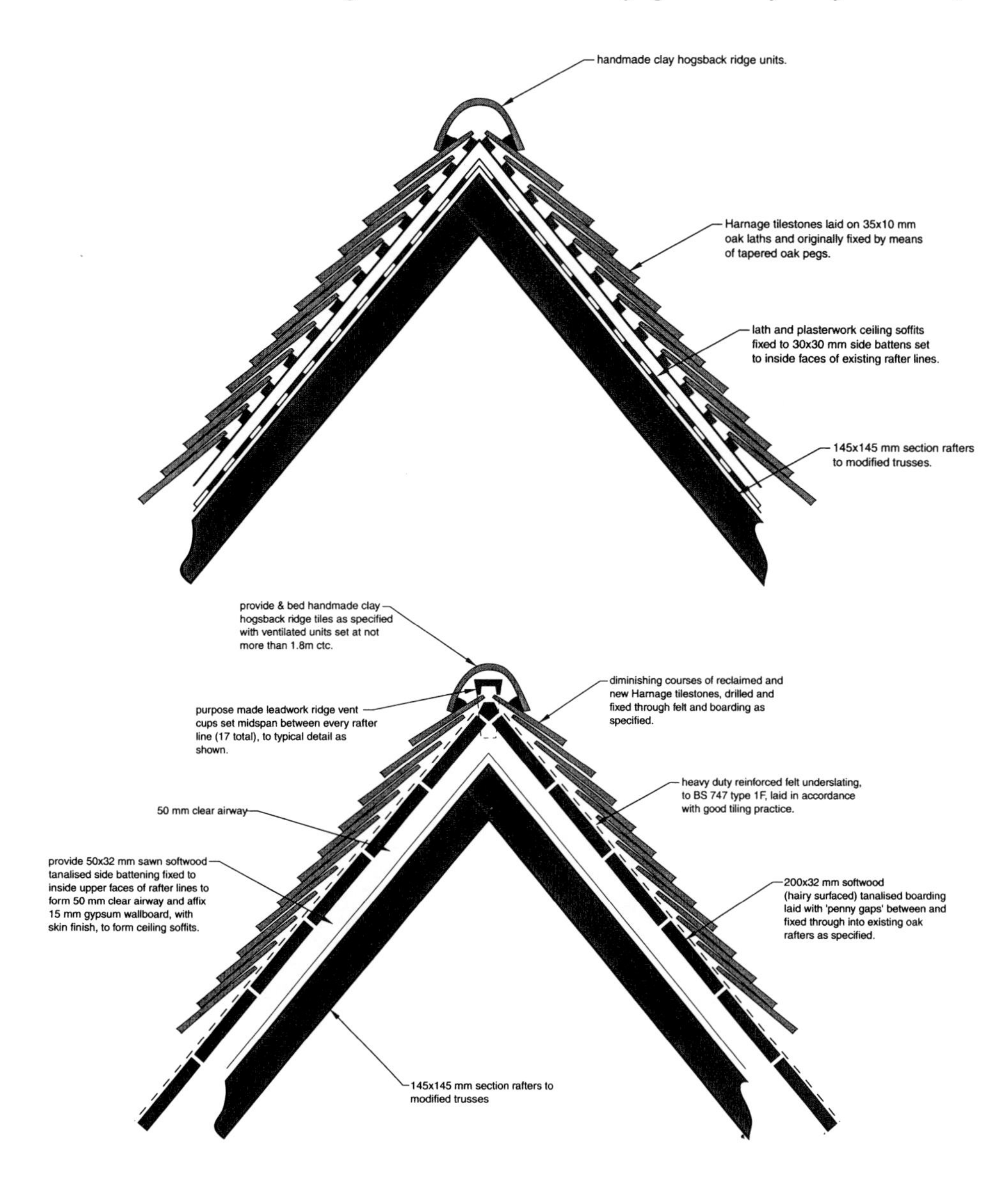

Figure 30. Typical section showing existing details (from original drawing, © Andrew Arroll).

Figure 31. Section showing proposed details (from original drawing, © Andrew Arroll).

which would improve the waterproofing around the nails.

Providing the eaves ventilation meant cutting out the rendered common bricks (possibly inserted during the 1910 works) and building in 28 terracotta air bricks, 215 mm wide x 65 mm high (9" x 2½") between the rafters with a nylon mesh behind to stop insects (Fig 32). The ventilation would flow through each rafter space between the boarding and the ceiling and would be vented at the ridge via hogsback ventilation ridge tiles. Ten airbricks were also used to ventilate the body of the church at wall-head level.

To comply with English Nature's recommendation, the specification stipulated that the slates should be stripped as early in the year as possible, before the bat breeding season, to discourage them from returning to the bellcote.[9] A comprehensive method for assessing the suitability of the slates for reuse was included. As the roof was stripped, each slate was to be individually assessed by a competent craftsman roofer using three tests.

The first was to assess strength and durability, looking for cracks, splits, fissures, de-lamination and signs of crumbly or soft sandy texture. The sound slates should be dense, hard, free from splits and shakes and when tapped sharply, give a true ring without any indication of splitting or delamination. The sides should be relatively straight and the whole slate reasonably rectangular. The second test was to select only slates with a minimum width of 200 mm (8") to ensure a minimum of 100 mm (4") side lap. Thirdly, rejected

Figure 32. Terracotta air bricks which replaced common bricks at eaves level. These are intended to ventilate through each rafter space between the boarding and ceiling and out via the ridge vents (© Chris Wood, English Heritage).

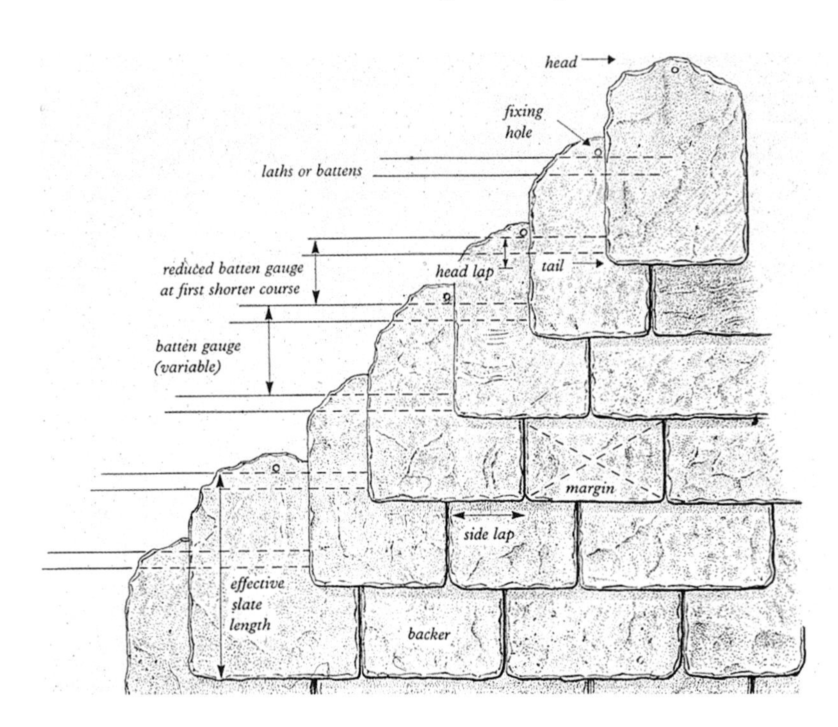

Figure 33. Diagram showing how slates are measured (1997, © English Heritage Photo Library).

slates that were sound were to be reserved for redressing to smaller size.

Following stripping a thorough examination of the condition of timbers was to be made. The details for repairing upper rafter surfaces was to screw and glue on new moisture-matched oak to prepared surfaces, probably two laminations.

English Heritage agreed that shouldering on slates would be limited to 25% of the length and width. The minimum headlap for the slating was specified at 100 mm (4") for the largest slates, reducing to 75 mm (3") and 50 mm (2"), but not more than one third of the slate length, in the middle and upper courses. Normal increases of headlap to accommodate reductions of slate length would, of course, apply. The slating would be set out to achieve as large a sidelap as possible, that is, slates would be chosen with the appropriate width to place the perpendicular joint as centrally as possible in the slate below (Fig 33). This is normal practice. But, taking into account the problems with the previous roof, to ensure sufficient sidelap the minimum for the middle courses was specified at 75 mm (3") and, for the last few courses, 50 mm (2").

Mortar was to be restricted to small dabs to prevent rocking[10]; it was not to allow wicking to take place up to the head of the slates. During the stripping, notes would be made of existing courses, with the difficulties in alignments and variations in existing sprocket dimensions, angles and setting out resulting from the irregular plan geometry of the church walls.

The contractor

English Heritage had recommended that a direct contract be awarded to an experienced firm of roofing contractors familiar with Harnage stone. However, because the job included other building works and for reasons of contractual expediency, the PCC and its architects wanted to appoint a main contractor who would include an experienced roofing firm within their tender. Five experienced conservation contractors were invited to inspect and submit a method statement together with their tender. The tender was awarded to T J Crump Building Conservation Ltd, Hereford, who included Heritage Roofing, Hereford, as the specialist sub-contractor. They were currently working on five re-roofing projects using random stone in diminishing courses and were interviewed by A&S prior to appointment to satisfy themselves that they had the skill and judgement needed to implement the work. They proposed to use five experienced roofers, all time-served craftsmen.

Their method statement included the following:

- numbering each hip cut from the bellcote and marking with chalk prior to removing slates
- recording of the roof by photography and a written record of the number and position of slate courses
- lifting and cleaning of ridges
- inspection of each slate using the three specified tests: stripping the top third on one side, then the other and progressing down alternately to avoid placing an unbalanced load on the roof
- leaving battens for photography and measuring
- counting all slates
- nailing experiments to be carried out prior to felting
- after roof boarding, installation of two layers of underlay
- preparation of a battening schedule, marked in chalk on the underlay and reslating to be carried out in alternate sections to avoid uneven loading
- bedding ridges
- implementation of the same method for the bellcote.

Stripping of roofs

Not surprisingly, the initial stripping of the roofs revealed that wet rot (believed to be a Basidiomycete wet rot) had affected the laths and to a lesser extent some of the rafters to a depth of 30–40 mm (1¼-1¾"). This was due mainly to the entrapment of moisture caused by the heavy mortaring. Isolated pockets of deathwatch beetle *(Xestobium rufovillosum)* had affected some small areas of the wall plates where mortar had been daubed over in the past. This had not weakened the wall plates. Most rafters, tilting fillets and other secondary elements had survived quite well.

THE RE-ROOFING

Site tests

In early March 1998 site trials were carried out. The roof boarding samples were delivered, but they had been cut continuously vertically, plank after plank, and had to be replaced. Each plank would have had a lot of sapwood which would have led to cupping or bending on drying. The correct way is to saw the log into quarters and cut the planks mainly from the heartwood so that any distortion is minimized. To avoid trapped condensation between the upper surfaces of the boards and the underside of the felt/separating membrane, it was agreed that a 10–12 mm (½") gap between boards should replace the originally intended 5 mm (¼") gap. This would allow for any dimensional change and each board could in effect be regarded as a large batten. The boards would not therefore act as a layer of thermal insulation above which condensation could occur and be trapped. Some condensation could occur at the interface of the felt and the boards but this was expected to drain away through the gaps, which would also allow partial ventilation across the rough sawn upper surfaces of each board and underneath the felt.

The merits and drawbacks of installing a vapour barrier, as recommended by BS 5250:1989 *Code of Practice for Control of Condensation in Buildings*, were debated at length. Ideally this should be located on the warm side, but this was not feasible as it would be on the lower side on the plaster ceiling. Other positions were limited especially as the boarding could no longer be regarded as a layer of insulation because of their spacing. Condensation would occasionally occur, for example, against the underside of the sarking felt when there was a rapid heat loss to clear skies. But this should clear by ventilation within the rafter space. However, any condensation against a vapour barrier would tend to stain the white ceiling immediately below, so the vapour barrier was excluded.

The decision had been made to board horizontally with nailing kept five slate nail diameters away from the edges of individual boards. Consequently, the boards had to be sawn to varying widths gauged out to the slating.

A sample of the largest of the reclaimed slates was nail-fixed using 4.5 mm (¼") diameter stainless steel ring-shanked nails in the two existing 10 mm (½") diameter holes in the slate. Although the nails held the slate under its own weight only a small downwards pull was needed to bend the nails until the head jammed diagonally across the oversized holes in the slate. Also, because the undersides of Harnage slates are irregular, there can be up to 20 mm (1") of unsupported cantilevered nail shank between the upper surface of the boarding and the underside of the slates (Fig 34). This, too, tended to bend the nails so it was agreed that plastic sleeving would be used around the fixing nails to hold them tightly in the old, large holes. Holes in new slates would be drilled to fit the nail diameter. Because stainless nails were not available with large heads, stainless steel washers were used. With smaller, thinner slates further up the roof, bending problems would not occur. The nail lengths were varied to suit the thickness and flatness of the slates and the tilt at the eaves to ensure adequate penetration into the boards.

A trial panel of slating was carried out to confirm that these procedures, materials and proposed laps were practical and would prevent rain penetration down to sarking felt level.

To prevent any uneven slates rocking, the use of very small spot mortar bedding was permitted. This was to be kept well away from slate edges. A very stiff mortar mix was suggested using a moderately hydraulic lime [11] and well graded aggregate (in the proportions of 1:2½ binder: aggregate by volume) with some hair to discourage later fragmentation.

The works

Stripping the slates began in early March 1999 and by 11 March enough had been uncovered to show the extensive damage to the laths. Considerable timber decay had also occurred at the abutment with the clay tile roof due to inadequate lead soakers, so the design was modified to carry the soakers up and over a secret gutter under the verge of the stone-slating.

Slates were inspected for reuse and stacked on the working scaffold and on the ground. As the numbers and sizes of usable slates became clear, CSC were instructed to ensure that sufficient numbers of slates of each length would be available. CSC proposed delivering slates in sufficient numbers to satisfy a theoretical gauging and setting out of the slating, taking into account the numbers and sizes available for reuse. Helpful as this was both in

Figure 34. The stainless steel nails and washers have a plastic sleeve on the 20 mm of unsupported shank below the slate. Without the sleeves the nails could bend within the oversized holes from the weight of the slate (© Chris Wood, English Heritage).

Figure 35. A mock-up of how lead soakers acting as 'shadows' can overcome the problem of marginal laps (© Chris Wood, English Heritage).

controlling costs and facilitating progress, it was to lead to a problem, discussed below.

By this time the replacement, correctly-sawn, boards had been delivered to site. The final decision was taken to board the roof horizontally and to cut the boards to suit the slate gauging. The proposal to ventilate the roof and the body of the church was also reviewed and the details agreed prior to the boards being installed. Slating began on the north slope and initially was satisfactory but some problems were identified. Some of the slates were unacceptably shouldered, making it difficult to achieve satisfactory laps no matter how carefully the slates were selected. It was now becoming apparent that there was something of a mismatch between the way the slates were being made and the experience of the slaters. This sort of difficulty had been anticipated and showed why English Heritage had recommended that the slater be involved in the production of the stone slates.

Put simply, there are two traditions of stone slates:

- the large, flat and generally square, but modestly shouldered, sandstones typified by the Carboniferous stones-slates of the Pennines and Welsh Marches
- the 'Cotswold' type, which are frequently twisted, uneven, heavily shouldered and reduce to much smaller sizes than the sandstones.

The properties of Harnage slates are far more like the 'Cotswold' type than the 'Pennine', and CSC were using the former practice in their dressing, assuming that the use of Cotswold techniques would be followed by the contractors. Additionally, this meant that to ensure satisfactory laps where shouldering was doubtful, 'shadows' would be used. These are small pieces of thin stone slate set into mortar over a bad shoulder to weather the lap: normal practice in the Cotswolds. The technique was unfamiliar to Heritage Roofing's slaters. In fact, to overcome this problem, two decisions were taken. Firstly, the worst examples were to be re-dressed to improve the shape at the head and, secondly, where the adequacy of the laps at the shoulders were doubtful they would be improved by the use of shadows. But, because suitably thin pieces of Harnage stone were not available, they would have to be made of Code 4 lead (1.8 mm thick, from BS 1178:1982 *Milled Lead Sheet for Building Purposes*), and applied as soakers (Fig 35).

The first two courses were approved but as work progressed the slaters ran into another problem: they had accepted the calculated gauging provided by CSC to set out the roof. This was correct for the slates supplied but because there was only exactly the correct run (ie total width) of slates in each course there were not enough spares from which to choose slates wide enough to provide adequate sidelaps. Rather than struggle with this problem and risk leaks, Heritage Roofing took down their work and re-gauged the roof allocating sufficient spare slates in the lower courses to achieve satisfactory sidelaps. This was not easy because the roof boarding could not be changed and so the slating had to be gauged out to suit the boarding as well as the slate lengths and laps. Thc lack of spare slates would not be a problem once the first few courses were completed because the extra slates from the lower courses could be used higher up.

The thickness and unevenness of the slates did not pose a problem and, provided they were selected so that each slate 'nested' into the space in the course below, there was no need to use mortar spots. Thin slates were reserved for the shallow gable abutment where they had to fit under a stone coping.

The process of decision-making and tightening up the standard of work had been time-consuming and expensive, requiring many of the team to travel long distances to attend meetings. Considerable time was spent convincing PCC members about the methods adopted, as they were very anxious to avoid a repetition of the previous roof failure. Work was also slowed because many of the new slates still had silt and clay on them which had to be wire-brushed to discourage plant growth. Pressure washing was not possible because of the lack of an adequate water supply.

When work moved to the other side of the building, where the roof was out of square, the bottom courses were set out using longer slates at the splayed end, gradually adjusting the lengths in subsequent courses until the slating became parallel. This ensured that the correct head lap was maintained over the awkward shape of the building. This is an example of how skilled slaters, who clearly understand how slating works, can overcome most problems that the roof presents. Slating began again immediately but in the subsequent weeks progress was slow as great care was being taken to satisfy all the laying criteria. Additionally, the wire brushing and the inclusion of shadows were adding unanticipated costs and time. So it was becoming clear that the three weeks originally envisaged to slate the roof was an underestimation.

A month later 60% of the slating had been completed on the north pitch and 75% on the south but work had become very difficult (Fig 36). In the absence of battens to work from, intermediate platforms should have been provided by T J Crump, the main contractor. Instead, the slaters were having to work from ladders, meaning that slates had to be selected on the scaffold and passed up to

Figure 36. The attractive appearance of Harnage stone, with the roof now three-quarters completed (© English Heritage Photo Library).

the slater, a laborious process which mitigated against choosing the best slate for each position. Having done their best under difficult circumstances, when subsequently viewed from below they would find that the work had to be redone.

In the next month work progressed even more slowly but by the beginning of August the slating was within a few courses of the ridge. Then the reason for the slow progress became apparent when T J Crump, the main contractor, went into liquidation. Understandably, during the previous months Heritage Roofing had had to employ their men on other contracts to ensure a continuity of income. A further lengthy delay ensued while the contract was determined and a new contractor appointed and even though Heritage Roofing agreed to complete the work in spite of it now being a total loss to them (since T J Crump no longer existed to pay them), work could not start again until late October.

A final unanticipated difficulty arose at the ridge. Because the rafter space was to be ventilated at the ridge through lead 'chimneys' underneath the vent ridge tiles, the boarding had had to be cut back around the 'chimneys'. This meant that, in these positions, there was nowhere to nail the final two courses of slates. This was overcome by using wide slates to span the gaps. Fortunately, there were plenty of small slates to choose from. By the middle of November the bellcote slating had also been completed and on 19 November the work was accepted and the order given to strike the scaffolding (Fig 37).

A supply of new slates of different sizes was left with the church for any repairs that might be needed. The decision was made not to provide copper strips at the ridge as a way of minimizing organic growth. There was some concern about the appearance of the copper and its long-term effectiveness and whether it would stain the slates. In any event, a complete absence of all growth

Figure 37. The completed re-roofing (English Heritage Photo Library © Paul Highnam).

including lichen would not be desirable, so it was agreed that occasional maintenance would be carried out from ladders to remove any build-up of moss or larger plants. This would be carried out with the use of scrapers or suitable herbicides dispensed from long lances and hand-held pressure sprays.

COSTS

The total cost of the whole works included repairs to the walls and interior of the church as well as improvements to drainage. These were carried out by T J Crump, the main contractors, who would then pay Heritage Roofing. The total figure included fees, insurance and archaeological recording.

The geological study involved desk and field research and was estimated to cost £1,200. The trial excavations that took place at the two sites cost £1,250 and this included the hire of the equipment. Both Terry Hughes and Dr Jefferson were consultants and already under measured term contracts to English Heritage. Some of their costs were absorbed under these contracts and not merely paid out of the grant, so these figures may be slightly under-estimated.

Four days were allowed for Terry Hughes's roof investigation, which cost £2,000 and included a labourer, lighting and incidentals as well as the preparation of his report. It did not include the scaffolding cost.

The total cost of scaffolding was nearly £18,000 which included the investigatory scaffold. The delays in the contract increased the expected costs.

The cost of extracting the stone, including setting up and reinstatement was £15,500 for 27 days on site. The total cost of production, which included transport to and from Wiltshire, plus the dressing and preparation of slates was £44,600 (below budget). Including the stone for Pitchford Hall, approximately 90 tonnes of stone was taken down to Wiltshire and, allowing for an average wastage of 40%, resulted in 62 tonnes of dressed slates. This meant that a tonne of slate ready for use cost approximately £620.

Working on the approximate basis of 2 tonnes per square, the cost was £1,240 per square. This can be compared with the nearest available equivalents at the time, which were £1,100 for 'Cotswolds' or £1,050 for French limestone (although this had a higher wastage factor). Clearly Harnage was more expensive than the alternatives. However these other producers were able to build up economies of scale based on continuous production, something which could not be done on a 'one-off' project like this. Given all the scale factors and the difficult working conditions the final cost was considered to be very reasonable.

CONCLUSIONS

The main objective of this project was to secure the re-roofing of Pitchford Church by re-using existing Harnage slates and making up with new, in line with English Heritage's policy to retain and foster stone slate roofing. This was successfully achieved and there were other benefits including additional supplies of slates for the church for maintenance purposes, for Pitchford Hall and for English Heritage's own estate and other grant-aided work. Some of the smaller slates have already been used to make up shortfalls during repairs to Cressage Old Hall 6.5 km (4 miles) to the east and Berrington Church, 3 km (2 miles) to the north. If new Harnage slates are needed in the future, adequate supplies have been identified and acceptable methods of temporary exploitation and production established.

Lessons have been learned which could be of value to those in other stone slating areas wanting to win stone for use on roofs. It must be remembered that this project was an unusual prototype involving a large element of research and development resulting in extra costs. However, this exercise has provided a great deal of knowledge and experience of small-scale temporary delving. In this part of Shropshire, the stone had not been commercially exploited for over a century and the number of roofs still clad in Harnage slates is extremely small. This very limited market meant that it was difficult to establish economies of scale, by producing large surpluses for sale to others. The unit cost of the slates was therefore high, some 13% more than the nearest 'similar' and available stone slate. However this would have been limestone slate imported from the Cotswolds, so it is questionable whether it could be described as similar. Harnage stone also costs more to lay than other more uniform slates because of the greater time taken to sort and select material which is very uneven and variable in thickness. All stone slates are variable in size.

The value in carrying out proper research, both documentary and practical, was clear. Some of the nineteenth-century geological maps were vital in identifying the most likely quarry sites and the papers detailing the repairs carried out in the 1930s give an insight into why the roof failed prematurely. It was clear that the failure to obtain an adequate number of suitably sized new slates meant that the roofers had to use an alien technique (mortar bedding). This research also suggests that despite the expertise and ability of the contractors, the decision not to use an experienced architect in the 1930s might well have been a false economy.

Practical investigation also yielded invaluable information. The field trials of potential delves were vital in ensuring that slates of the right size and type could be produced. It also established the extent of the potential resource. Ideally, the existing roof should have been investigated before delving, so that a product specification could be developed. This could have included the type of stone, the thickness, flatness, shape and size range and the quantity needed to complete the works. This would have prevented some of the problems which were encountered on the roof. Nonetheless, the investigative survey showed clearly why the roof had failed and confirmed many of the faults that had been anticipated. Such an investigation does require the erection of a sturdy protected scaffold and the relevant expertise in carrying out this work. Even if a comprehensive survey such as this

Figure 38. The site at the height of production (© Chris Wood, English Heritage).

Figure 39. The same view after restoration of this historic site (© Chris Wood, English Heritage).

is not possible, some form of historical recording is important if knowledge is to be gained of earlier detailing.

Much can be learned from experienced tradesmen, particularly those familiar with traditional local practice. In this case no-one was found, but it must be borne in mind that one slater's current practice may not reflect the historical norm. Advice was given by a former local quarryman at the delves about the quality of rock that could be won and its ability to dress down to the required sizes. Expertise from professionals and tradesmen was helpful in preparing the trials on the roof and the specification.

Obtaining permissions and consents can take a long time and much abortive work was avoided by constructive dialogue at a very early stage. This involved a number of parties, principally land-owners and planning authorities. Church of England faculties were also required for the changes to the church and an English Nature bat license was needed to work on the bellcote. With so many people and organisations involved it is essential that project management is tightly controlled, preferably by one person in sole charge. The main recommendation to anyone contemplating a similar exercise is that a project co-ordinator should be appointed right at the start, with responsibility, resources and authority to ensure that the project is not delayed at each step, at least until management of the works is taken over by the architect.

The project involved a large team (listed in Appendix 1) which needed clear lines of communication and decision-making, particularly once work began on site. Generally this was achieved with A&S, acting for the PCC, making the final decisions. The method statement was extremely important in establishing much of the working practice, but it is important to remember with random slating that many decisions need to be taken by the craftsman on site and that adequate reserves of all sizes of slates are needed to give them the flexibility to construct a water-tight roof.

POST SCRIPT

In summer 2000, the Acton Burnell Estate asked CSC to reinstate the delve. This work was completed in early autumn and Figures 38 and 39 show the site during working and after reinstatement. Should there be a demand for future supplies of Harnage stone, then, subject to the various consents, the Park Wood delve could be opened again, relatively quickly.

ANNEX 1: THE TWENTIETH-CENTURY WORK ON THE CHURCH

1910 Works

A 'restoration' of the church was carried out between 1908 and 1910, although many of these works appeared to be essential repairs. In June 1908 the incumbent, Prebendary Machen, wrote:

> The church has for many years been in a state of dirt, dilapidation and decay. The woodwork is rotten and worm-eaten, especially at the bottom of the pews and the floors of which rest on the bare soil and which are full of holes. The font is broken and full of dilapidated hymn books: it has no drain, no lining and no cover. I am informed that a 'slop basin' has been used for the Holy Baptism in the past. The altar is totally bare, except for a superfrontal, no Cross no lights. The Sanctuary is covered with old and very dirty coconut matting. The Heating Apparatus consists of four holes in the aisle covered by iron plates, in each of which a separate fire is lighted, filling the church with smoke and fumes, and being distinctly dangerous to walk over. The walls are dirty and the plaster is cracked and also full of nails of all sorts and sizes, put there presumably to fasten 'decorations'. The plaster has been patched in many places, and a different shade of colour seems to have been used each time. The timber roof is entirely hidden by a plaster ceiling and there is a plaster cornice on each side. The service books are very old and dilapidated. The vestry is full of rubbish. There is no choir and no choir books. (Machen 1908)

Various works were put in hand, but the church was not closed for the restoration until Easter Monday March 28th, 1910, because of difficulties obtaining patron's consent.

> The longest part of the work was doing the roof. The plaster ceiling was knocked down with iron bars by men standing on the rafters above. When this was finished, the rafters were cleaned down and repaired where necessary. The oak trees from which the new tiebeams and king posts were made were felled on the Pitchford estate and cut out on the circular saw belonging to Colonel Cotes, they were then (when completed in the rough) carted down to the church where the joiners finished them. Hoisting the tie beams into place was a long and difficult business but it was accomplished without any accident of any kind. When these beams and posts were in place the whole timber roof was stained to match, also the new oak cornice. The whole floor of the church was taken up and a concrete foundation put in. The new heating apparatus was erected and was found to work satisfactorily. (Machen 1908).

The remainder of the works described were to fixtures and fittings. No description was given of the recovering of the chancel or nave roofs. Presumably the stone slates must have been stripped and re-used on the nave. It was at this time that the chancel was covered in the Broseley tiles, possibly because there was not enough salvaged, usable stone to recover the whole roof.

Condition of roof in the 1920s

Despite these works the nave roof still presented a problem. 'The dilapidated and dangerous condition of the roof was known as far back as 1925, but the village in spite of Prebendary Machen's reports, elected to build a village hall before repairing the church' (Stokes 1937). In 1924 a Mr H Gibbon reported that 'rafters are strained to breaking point and are 1'3" hollow ... weight of stone tile roof beginning to tell ... The rafters have no purlins to support them' (Gibbon 1924). Mr E L Wratten (of Wratten & Godfrey) wrote that the whole of the roof was 'almost entirely covered with thick moss ... many slates have slipped and others become so irregular that they cannot be relied upon for keeping out heavy rain. On the north side the rafters have sagged very noticeably and at some previous date pieces of timber have been inserted in order to level up the backs of the timbers and enable the slates to lie upon a reasonably true surface' (Wratten 1924). With the exception of the new pieces which had become saturated with moisture and decayed, Wratten considered that it was in no way different from its condition in 1910 when some repairs were carried out. Although the main structure of the nave roof was replaced in 1910, it seems clear that the stone slates were still bearing on the decayed rafters.

In 1927 Rev Stokes instituted moves towards remedying the situation by inviting an architect of the Lichfield Diocese (Mr Thompson of Derby) to investigate, but after a very casual survey he concluded that the roof might last another 50 years. 'A fund ought to be started in case of accidents and tell-tales placed in the walls' (Dixon 1936). In the event roofing work did not commence for another nine years.

Roof works in 1936

Damp was getting in by 1935 and Mr B Treasure (Treasure & Son, Building Contractors, Shrewsbury), without the benefit of a detailed investigation, thought he could relath and re-slate for about £100 (Dixon 1936). The following year General Sir Charles Grant accompanied Rev Stokes to a PCC meeting and persuaded them to put the work in the hands of Treasure & Sons.

Treasure's report (Treasure & Son Ltd 1937) describes that on July 1st 1936 they started work on the nave roof. At the time it was hoped it would only be necessary to strip off the slates, replace the decayed laths, and do a certain amount of repairs to the woodwork, refix the old slates, making up with new quarried ones when required. On 28th July Mr E Fox, presumably of Treasure's, wrote of his concern about the oak roof. Taking out any of the principal rafters would have been hopeless as they were out of:

> cloven timber of irregular shape and have cambered by 12". Some of the rafter tops are completely gone ... the rafters or principals are out of plumb quite 12" ... The present state of the timbers is not good and should it be decided to leave the roof on then I cannot accept any responsibility in a gale ... I am very definite in stating that we are not going to make a satisfactory job by cobbling up the existing roof ... If possible the old roof should be cleared off and a new one in oak with old fashioned principals, purlins and oak rafters square pegged to keep character. (Fox 1936)

When the roof was stripped, the timbers were found to be in a very bad state, apparently being quite inadequate to carry the weight then imposed upon them, deathwatch beetle and 'small oak grub' being responsible for 'the roof having been defective for years past'. Rev Stokes reported that as the shingles were removed 'it became apparent that damp, rot, and the deathwatch beetle had done irreparable damage to the whole fabric of the roof (Fox 1936). The decision was accordingly taken, supported by the Diocesan Building Society, to reconstruct the entire roof as an exact replica of the old one. It was decided to replace the existing timbers with new, using only the existing tie beams and king posts and internal cornices (1910 work). Interestingly, they also replaced the timbers supporting the chancel roof which had been recovered in the Broseley tiles.

Although not anticipated at the time, it was necessary to quarry a considerable number of new stone slates to make up the defective ones taken from the roof and also to make up the loss in size of old ones, which in most cases had to be redressed, shaped and reholed. The quarrying on Acton Burnell Hill (OS SJ 538007) was slow and costly, as 'the beds of stone lay in very awkward positions, and the suitable stones were few in number, and necessitated a great amount of clearing of top material to arrive at suitable stone' (Fox 1936). About 75% of the old stone shingles were found unfit to be used again. 'This proved painfully costly, and our expenditure rose by steps from £600 to £800 and then to £1200' (Fox 1936).

All woodwork, apart from tie-beams, king-posts and internal cornice, had to be renewed, apparently sound timbers having been affected by deathwatch beetle. 'The timbering of the chancel roof had to be made good, but the Broseley tiles on the chancel, which had been substituted for stone shingles [ie stone slates] in Preb. Machen's time were put back, for the additional cost of more stone shingles could not be faced' (Fox 1936). The whole of the internal plastered ceiling was renewed and the walls made good with patch repairs. The timbers of the belfry were also in poor condition, but these were not replaced as it was possible to cut away affected surfaces and treat them with an unspecified insecticide and preservative. Some weakened timbers were strengthened by splicing on pieces of best timber taken from the roof. Cracked stonework was made good, buttresses pointed and grouted and all guttering made good, downpipes and drains cleaned out.

An appeal for funds was launched, which estimated the cost of repairs to be £600, and this apparently brought responses from many parts of the world. Contributors included Treasure's masons 'who voluntarily made a small contribution from their wages, every week, towards our costs' (Fox 1936). The appeal leaflet was also sent to the secretary of the Society for the Protection of Ancient Buildings (SPAB) who responded by requesting the name of the architect advising on the works. Rev Stokes replied 'as the work on the church was simply a replacement of roof, an exact copy of that which was there before, General Sir Charles Grant and I sought the guidance and oversight of the Hereford Diocesan Church Building Society (the Diocesan Advisory Committee had previously been consulted), of which Dr W Watkins Pitchford of Bridgnorth is Chairman. This Society agreed to our putting the work in hands of Mr Bernard Treasure & Sons, who have done much of this kind of work … I have been struggling for 10 years to get this work begun and money is too scarce to pay Architects' fees, where it is not absolutely necessary' (Fox 1936).

The SPAB's secretary replied to the Archdeacon of Ludlow stating that the 'Society has little information as to the value of the church, or of the works in progress, but a letter from the Vicar, which ... is disturbing'. The Archdeacon replied that he had made enquiries and confirmed that the Vicar had inspected the work on several occasions. 'He says that the firm which is doing the job is an excellent firm. I myself can confirm this and that the quality of the material which is being put in, is good. He thinks that the work being straight-forward, there is no need of an architect'. The Committee of SPAB sought clarification as to whether the roof being replaced was original or a later restoration and whether it was condemned solely on the report of the builders. The Archdeacon sent on the Vicar's correspondence and also confirmed that Dr Watkins Pitchford had seen the work and confirmed that in his opinion the work was being well done and good material was being used. From the papers, the Vicar concluded that it was a sixteenth-century roof, 'patched up about 60 years ago when new oak tie beams were introduced to prevent the walls spreading, work abominably done and casing in the beetle grub, which must have been then present' (SPAB 1936).

The SPAB's motive for pursuing the issue was presumably their desire to see the roof repaired with the absolute minimum loss of historic fabric. Good independent professional advice may well have confirmed that it was indeed necessary to cut out and replace much of the structure. However a competent architect would also have ensured that the new roof was detailed and constructed correctly and this aspect seems to have been ignored. 'I may have been wrong in not calling in an architect but as the work was simply a faithful copy of the roof as it was before, I could not see the need for that expense' (SPAB 1936). Unfortunately, the design of the new roof either perpetuated faults in the old roof or failed to replicate the original design, but this did not become apparent for another 50 years.

Condition of the nave roof by 1995

The condition of the nave roof was a clear cause for concern by 1995. The report of English Heritage's commissioned architect, John Wheatley, to the Churches' Sub-Committee of its Historic Buildings Advisory Committee in October 1995 included a description of condition of the nave roof. 'Both pitches in Harnage fossiliferous limestone slates of local origin. These renewed/relaid most recently 1937. Unfortunately full bedded on hard, probably cement mortar. Thus both pitches monolithic. Widespread failure of battens by rot caused by the embedment in and torching from below of cement mortar. On N pitch, sections of slating folding inwards between rafters, displacing lath and plaster work beneath, and allowing ingress of water. Mortar prevents lifting of slates to inspect work beneath. In probability no useful reclaim of slates may be possible. N slope heavily vegetated and mossed. Although in terminal disintegration, diminished course slating retains attractive and distinctive appearance matching most roof pitches of adjacent Pitchford Hall' (Wheatley 1995). A grant of £31,414 (60%) of eligible costs was offered but this was revised in January 1998. Based on assumptions about the costs and extent of works a new grant of £76,198 was offered, which amounted to 80% of the costs of renewing the ceiling, re-roofing the nave and repairs to rainwater goods and drains. The increased amount offered reflected the escalating costs attributable to English Heritage's requirements.

ANNEX 2: THOSE INVOLVED WITH THE 1998–99 PROJECT

English Heritage

Chris Wood, Building Conservation & Research Team, English Heritage

Ian Forrest, quantity surveyor, English Heritage

John Wheatley, inspecting commissioned architect, Wheatley Lines, Brickfield, Bickley, Tenbury Wells, Worcs WR15 8LU, UK.

Consultants

Terry Hughes, Slate & Stone Consultants, Ceunant, Cernarfon, Gwynedd LL55 4SA, UK.

Dr David Jefferson, Jefferson Consulting Ltd, The Old Armoury, Unit 13, Orb Way, Crown Business Park, Station Road, Old Dalby, Melton Mowbray, Leicestershire LE14 3NQ, UK.

Parochial Church Council

Andrew Arrol, Arrol & Snell Ltd, Architects, St Mary's Hall, St Mary's Court, Shrewsbury SY1 1EG, UK.

Wilf Jones, quantity surveyor, John Pidgeon Partnership, Suite 5 Noram House, Victoria Road, Shifnal, Shropshire TF11 8AG, UK.

Rev Frank Rumball, The Rectory, Condover, Shrewsbury SY5 7AA, UK.

Alex Argyropulo, churchwarden and project liaison, Coachman's Cottage, Pitchford, Shrewsbury SY5 7DP, UK.

Richard Perry, PCC Treasurer, The Old Rectory, Pitchford, Shrewsbury SY5 7DP, UK.

Acton Burnell Estate Trustees

Mrs S Baker, Balfour Burd & Benson (formerly Cooke & Balfour), 1-2 School Gardens, Shrewsbury SY1 2AL, UK.

Pitchford Hall Estate

Andreas Letch, Estate Manager, Pitchford Hall, Pitchford, Shrewsbury SY5 7DW, UK.

Shrewsbury and Atcham Borough Council

Micky King, conservation officer, Shrewsbury & Atcham Borough Council, Oakleigh Manor, Belle Vue Road, Shrewsbury SY3 7NW, UK.

Shropshire County Council

Malcolm Bell, Minerals Planning Dept, Shropshire County Council, Shirehall, Shrewsbury SY2 6ND, UK.

Archaeology Unit, Shropshire County Council, Winston Churchill Building, The Radbrook Centre, Radbrook Road, Shrewsbury SY3 9DU, UK.

Principal contractor

T J Crump Building Conservation Ltd, The Lakes, Swainshill, Hereford HR4 7PU, UK (no longer trading).

I J Preece & Son Ltd, The Yard, Green Lane, off Upper Road, Meole Brace, Shrewsbury SY3 9JH, UK.

Domestic sub-contractors

Heritage Roofing, The Shrewd, King's Pyon, Hereford HR4 8PP, UK.

North West Lead, The Cottage, Barlowfield, London Road North, Poynton, Chester SK12 1BX, UK.

Scaffolding sub-contractor

South Shropshire Scaffolding Ltd, Bromfield Road, Ludlow SY8 1DN, UK.

Quarrying/delving contractor

Chris Harris, The Cotswold Stone Tile Company, Wedhampton Manor, Devizes, Wiltshire SN10 3QE, UK.

ENDNOTES

1. These were reclaimed stone slates mainly from the Old Red Sandstones from Herefordshire, but also included some limestone slates from the the Costwold hills.
2. These were mainly the geological and Ordnance Survey maps. OS maps of various scales from the mid nineteenth century to the present day were also studied. The archived file contained information from Murchison 1839, Pocock *et al* 1938, British Geological Survey 1850–80, British Geological Survey 1967, British Geological Survey 1978, Lawson 1985.
3. Chris Harris was actively involved in attempts to open new stone slate delves and quarries in different regions of England. He was keen to study the requirements identified for new material for Pitchford Church. His selection as the preferred contractor was made by the Acton Burnell Estate on whose land the delph was situated.
4. The Tetbury Stone Company Ltd, Veizeys Quarry, Avening Road, Tetbury, Gloucestershire GL8 8JT, UK; Tel: + 44 1666 503455.
5. The site forms part of the medieval deer park which was developed as a landscape park associated with Acton Burnell Hall built in the eighteenth century. Most of this is wooded apart from the area where the delph was to be sited. The parkland has a Grade II listing which means that the local authority has to take into account the effect on the parkland of any proposed development..
6. CSC's premises near Rowde, Wiltshire, comprised large, former agricultural buildings for indoor dressing of stone using hand and machine tools. This meant there was no disturbance to neighbours at Park Wood and allowed working to continue during winter months. Secure storage of the finished stone slates all sorted to size was another advantage.
7. The Church of England's faculty jurisdiction system covers works for which listed building consent would be required under the secular system; it also covers repairs and new contents which would not be subject to secular control. The Diocesan Chancellor decides the faculty application, taking advice from the Diocesan Advisory Committee. Changes to faculty jurisdiction (introduced by a new Care of Churches Measure 1991) introduced in 1993, require English Heritage, local authorities and the national amenity societies to be especially 'cited' (ie notified) on proposals affecting the character of a listed church or its archaeological importance, or on the demolition of an unlisted church in a conservation area, before the Chancellor decides a faculty application.
8. The strength of metamorphic slate is commonly anisotropic, ie stronger in one direction in the plane of cleavage than the other. To avoid failure on the roof, such roofing slates should be manufactured with the weakest direction orientated north-south. Slates which have the weakest direction east-west are termed cross-grained. Similarly, correctly manufactured slates should not be turned through 90° when fixed. Sedimentary stones do not exhibit this feature and may therefore be turned without detriment.
9. The Wildlife and Countryside Act 1981 gives full protection to bats because of their special requirements for roosting. It is illegal not only to intentionally kill, injure of handle any bat, but also intentionally to damage, destroy or obstruct access to any place that a bat uses for shelter or protection, or to disturb a bat while it is occupying such a place. The Act provides defences so that building, maintenance or remedial operations can be carried out in places used by bats. This requires early notification and discussion with English Nature. A&S duly discussed the issue with them. Although the species of resident bat was not identified, they had been seen flying into the bellcote although no evidence was found of their presence elsewhere in the church.

10 The mortar specification for the spot bedding comprised 2½ parts of lime putty: 1 part crushed brick dust: 9 parts of well graded grit

11 NHL (Natural Hydraulic Lime) 3.5 to BS DD ENV 459-1 was suggested. St Astier NHL 3.5 is supplied by Setra Marketing Ltd, 16 Cavendish Drive, Claygate, Surrey KT10 0QE, UK; Tel: + 44 800 783 9014; Fax: + 44 1372 801302; for technical enquiries and the names of local suppliers.

GLOSSARY

Bench: a working face and level in the delph/quarry

Bund: stored overburden

Facies: the sum total of features such as sedimentary rock type, mineral content, sedimentary structures, bedding characteristics, fossil content etc which characterize a sediment as having been deposited in a given environment

Rivings: the roughly split stones prior to dressing of surfaces and edges

Shadows: thin pieces of slate used in the Cotswolds and the Horsham district to improve the weather resistance of the roof when the head lap is judged to be inadequate for a variety of reasons. Originally the shadow was a thin piece of stone but it is now normally a metamorphic slate or strip of lead.

Shoulder (verb): to remove the top corners of a stone; (noun) the top corners of stone slates. Excessive shouldering can result in a leaking roof.

BIBLIOGRAPHY

Primary sources

Dixon H T, 1936 unpublished correspondence to the Archdeacon of Ludlow, sent to Society for the Protection of Ancient Buildings, London.

Fox E, 1936, unpublished letter to Rev Stokes dated 28 July 1936, sent to Society for the Protection of Ancient Buildings, London.

Gibbon H, 1924, unpublished report dated 31st March 1924, correspondence sent by Archdeacon of Ludlow to Society for the Protection of Ancient Buildings, London.

Machen R D, 1908, *Report together with account book and papers,* papers lodged with County Archivist, ref 3348/ch/39.

Society for the Protection of Ancient Buildings, 1936, unpublished letter to Rev Stokes held in SPAB library, London.

Stokes A, 1936 unpublished correspondence with the Society for the Protection of Ancient Buildings.

Stokes A, 1937 unpublished account of works, Shropshire Records and Research Centre, Shrewsbury.

Treasure & Son Ltd, 1937, unpublished report to Rev Stokes dated 25 May 1937, Shropshire Records & Research Centre, Shrewsbury.

Watkins Pitchford W, 1947 Address given to the Shrewsbury Historical Association at a visit to Pitchford on 25th October 1947, Shropshire Records & Research Centre, Shrewsbury.

Wheatley J, 1995, Report to the Churches Committee of the Historic Buildings Advisory Committee, English Heritage.

Wratten E L, 1924, unpublished report by Wratten & Godfrey dated 27th May 1924, sent by Archdeacon of Ludlow to Society for the Protection of Ancient Buildings, London (1936).

Secondary sources

British Geological Survey, 1850–80 Six-inch Geological Sheets, 42SW, 49NE, 50NW.

British Geological Survey, 1967 *Church Stretton, Sheet 166*, 1:50,000 scale Geological Map

British Geological Survey, 1978 *Shrewsbury, Sheet 152*, 1:50,000 scale Geological Map.

British Standards Institute, 1982 BS 1178:1982 *Milled Lead Sheet for Building Purposes*, London, BSI.

British Standards Institute, 1989 BS 5250 *Code of Practice for Control of Condensation in Buildings*, London, BSI.

Building Standards Institute, 1995 DD ENV 459-1 *Building Lime: Definitions, Specifications and Conformity Criteria*, London, BSI.

Cranage D H S, 1903 *An Architectural Account of the Churches of Shropshire.* Vol 6, Wellington, Hobson & Co. Also available at Shropshire Records and Research Unit, Castle Gate, Shrewsbury SY1 2AQ, UK, ref: P219/B/1/9–34.

English Heritage, 1997 *Stone Slate Roofing* technical advice note, London, English Heritage.

English Heritage, Peak District National Park Joint Planning Board, Derbyshire County Council & DETR, 1997 *The Roofs of England: A Celebration of England's Stone Slate Roofing Traditions*, travelling exhibition, available from the Stone Roofing Association, www.stoneroof.org.uk

Hughes T G, 1997a *The Grey Slates of the South Pennines: the report of a study into the potential to re-establish the roofing slate industry of the region*, 2 vols, London, English Heritage, The Peak District National Park Authority and Derbyshire County Council.

Hughes T G, 1997b, *Harnage Tilestones Shropshire – A survey of the quarries in the Hoar Edge Grit from which tilestones and other building materials have been obtained*, unpublished report for English Heritage.

Hughes T G, 1997c, *Harnage Tilestones – Site investigations of two potential sources of a traditional roofing material in the Hoar Edge Grit Shropshire*, unpublished report for English Heritage.

Hughes T G, 1999, *St Michael's & All Angels Church, Pitchford, Shropshire – Harnage Roof Investigations*, unpublished report for English Heritage.

Lawson J B, 1985 Harnage slates and other roofing materials in Shrewsbury and neighbourhood in the late medieval and early modern period, in *Transactions of the Shropshire Archaeological Society* **64,** 116–118, quoted in Victoria County History Salop viii 1968, 68 and Local Studies library, Shrewsbury, MS 2, f12.

Murchison R I, 1839 *The Silurian System*, London, John Murray.

Pocock R W, Whitehead T H, Wedd C B and Robertson T, 1938 *Shrewsbury District, including the Hanwood Coalfield (One-inch Geological Sheet 152 New Series). Memoir of the Geological Survey of Great Britain*, London, HMSO.

Plymley J, 1803 *General view of the Agriculture of the County of Shropshire,* London, Richard Phillips.

ADDRESSES

Equipment details

Machines can be hired by the day from plant hire firms. The ones used in this project were hired from Wildes Plant Hire, Main Road, Dorrington, Shropshire SY5 7JE; Tel: + 44 (0)1743 718777; Fax: + 44 (0)1743 718867.

Ruberoid Zylex can be obtained at most builder's merchants. It is produced by Ruberoid Building Products Ltd, 14 Tewin Road, Welwyn Garden City, Hertfordshire AL7 1BP, UK; Tel: + 44 1707 822222; Fax: + 44 1707 375060; email: sales@ruberoid.co.uk

AUTHOR BIOGRAPHIES

Chris Wood has worked in the Building Conservation & Research Team at English Heritage for the last nine years. The team specialises in dealing with the problems of deteriorating materials on historic structures and he is responsible for coordinating research programmes, providing specialist advice on remedial treatments, running training and outreach programmes and leading national campaigns. He took over the *Roofs of England* campaign in 1998 when he became involved with the Pitchford project. He managed the English Heritage practical training centre at Fort Brockhurst, which offered specialist practical training concentrating on the repair of masonry structures. Prior to joining English Heritage he was a director of an architectural practice specializing in the repair and refurbishment of historic buildings. This followed 12 years as a conservation officer with two local authorities.

Terry Hughes worked in the Welsh slate industry at Penrhyn quarry for 10 years. He also has experience of the slate industry in the USA, throughout Europe and in India. Since 1983 he has represented the British slate industry on the British Standards Institution's committees for roofing slates (as chairman), the code of practice for slating and tiling and the National Federation of Roofing Contractors' slating and tiling committee. He has also been the chairman of the European Standards (CEN) Committee for slate and stone roofing products for 14 years. Since 1993 he has been an independent consultant. As English Heritage's consultant on slate and stone roofing he has carried out studies into stone roofing throughout England and has advised on the conservation and construction of the roofs of many historic buildings. As part of his work to support the regeneration of the stone slate delving industry he established and chairs the Stone Roofing Association and the Stone Roof Working Group.